Cracking the USMLE Step 2 CK Q Book

First Edition

Paul Edward Kaloostian MD, Carolyn Louisa Kaloostian MD, MPH, Sean William Kaloostian MD

Copyright © 2012 by Paul Kaloostian MD, Sean Kaloostian MD, Carolyn Kaloostian MD/MPH

All rights reserved. No part of this publication may be reproduced, distributed, or transmitted in any form or by any means, including photocopying, recording, or other electronic or mechanical methods, without the prior written permission of the publisher, except in the case of brief quotations embodied in critical reviews and certain other noncommercial uses permitted by copyright law. For permission requests, email to the publisher, addressed "Attention: Permissions Coordinator," at the address below:

paulkaloostian@hotmail.com

Printed in the United States of America

First Printing, 2012

ISBN   978-1-105-52094-5

About the Authors

Paul Kaloostian: Matriculated through Thomas Haider Accelerated Biomedical Sciences B.S./B.A./M.D. Program with Undergraduate work at University of California Riverside and Medical School at David Geffen School of Medicine at UCLA. Currently, he is a Senior Resident at University of New Mexico Medical Center in the Department of Neurosurgery. He has a passion for treating the underserved and is fascinated by the immense cultural diversity in New Mexico. He is fluent in three languages: English, Spanish, and Armenian. He is an avid pianist and concert clarinetist and composes classical and Armenian Folk music.

Carolyn Kaloostian: Matriculated through Thomas Haider Accelerated Biomedical Sciences B.S./B.A./M.D. Program with Undergraduate work at University of California Riverside and Medical School at David Geffen School of Medicine at UCLA. Currently, she is a senior resident in the Department of Family Medicine at University of Southern California Medical Center. She is currently obtaining her MPH degree during her residency. She has a passion for treating the underserved populations. She is an avid ballerina having performed in multiple national performances. She is fluent in three languages: English, Spanish, and Armenian.

Sean Kaloostian: Matriculated through Thomas Haider Accelerated Biomedical Sciences B.S./B.A./M.D. Program with Undergraduate work at University of California Riverside and Medical School at David Geffen School of Medicine at UCLA. He is currently a resident at University of California at Irvine specializing in Neurosurgery. Of note, he is a National Rhodes Scholar Finalist and Varsity Baseball player. He has also completed many marathons with competitive times. He has a passion for treating the underserved communities. He is fluent in three languages: English, Spanish, and Armenian.

Editors

Paul Kaloostian MD- Chief Resident, University of New Mexico Department of Neurosurgery
Carolyn Kaloostian MD/MPH-Chief Resident, University of Southern California Department of Family Medicine and MPH University of California, Los Angeles
Sean Kaloostian MD-Resident, University of California at Irvine Department of Neurosurgery

Faculty Reviewers

Neal Bricker MD: Professor Emeritus Nephrology, University of California, Riverside
William Stringer MD: Physician, Professor of General Surgery, UCLA
Jim Colgan PhD: Professor of Anatomy and NeuroAnatomy, University of California, Riverside
Ameae Walker PhD: Professor of Histology, University of California, Riverside
William Kaloostian MD: Physician Kaiser Permanente Medical Center, Clinical Professor of Medicine at USC/UCLA
Aida Shirinian MD: Physician Kaiser Permanente Medical Center, Associate Clinical Professor of Family Medicine at UCLA
David Sklar MD-Physician, Dean and Professor of Emergency Medicine, University of New Mexico Medical Center

Contributing Authors

William Kaloostian MD
Aida Shirinian MD

Special Thanks/Dedications

This was a Herculean Task that could not have been possible without the efforts of many people. Special thanks to Eddie and Nanny for instilling with us the desire to give to others who are less fortunate. Thanks to Aida and Bill for their continued support and encouragement. They have dedicated their life to medicine and have always emphasized that knowledge is the key toward taking care of our patients. We also appreciate the immense effort of the Distinguished Nephrologist and Professor Neal Bricker for being a light for us in a seemingly long and dark tunnel.

Preface

The medical literature is enormous. It is filled with information that is growing as you are reading this sentence. Having entered the realm of medical school and residency, it is very difficult to gather and master the information that is most critical. After much review of current USMLE Board preparation texts, we have attempted to create a more concise and focused text with Question and Answer Book that addresses the heart of the major issues that are not only commonly tested but also encountered as one is taking care of sick patients. We have put much thought and effort into providing a text that can be used by pre-medical and medical students, as well as physicians and those in scientific fields. Our goal is to provide an avenue of knowledge that can be used to heal that which is most important to us: Our Patients! Enjoy!

-Paul, Carolyn, and Sean

We are always looking to improve this book! Please email any questions, comments, or concerns regarding the book and improvements that can be made to this edition to the following email address: paulkaloostian@hotmail.com.

Table of Contents

Chapter 1: Ethics
Chapter 2: Neurology
Chapter 3: Psychiatry
Chapter 4: Ophthalmology
Chapter 5: Cardiovascular
Chapter 6: Pulmonary
Chapter 7: Hematology Oncology
Chapter 8: Gastrointestinal
Chapter 9: Infectious Disease
Chapter 10: Musculoskeletal
Chapter 11: Rheumatology
Chapter 12: Endocrinology
Chapter 13: Renal/Genitourinary
Chapter 14: Electrolytes and Acid-Base Disturbances
Chapter 15: Obstetrics
Chapter 16: Gynecology
Chapter 17: Pediatrics
Chapter 18: Trauma and Post-Operative Management
Chapter 19: Dermatology
Chapter 20: Epidemiology

Hippocratic Oath

Introduction:

It is very difficult to master all the information that one is expected to know in preparation for the USMLE board examinations. We have attempted to author a text that is presented in a disease-based format, presented in a concise yet thorough fashion, with special attention to the high yield pathology often tested. We have included pictures that we feel is important to grasp the material presented in the text. Additionally, we have written associated Q books so that one may test themselves after much review of the text. We feel that it not the way that the questions are asked that determine difficulty but rather the problem lies within the knowledge of the individual. It is only through reading and re-reading this material that one may truly grasp the information. Once this is accomplished, then we feel that questions may be answered correctly, no matter how they may be written. In this text, we have tried to author questions in a manner that not only allows one to test their knowledge after reading the associated textbook but also to prepare for USMLE style questioning. Good Luck.

                                                                Paul, Carolyn, and Sean

Ethics

24-year-old female is involved in a car accident. Her vital signs are stable. She is awake and alert x 3 and following commands. She still has difficulty swallowing. Her physician wants her to have a g tube placed. She does not want a g tube. Her physician continues to tell her that she needs a g tube despite her unwillingness.

1) The physician is exerting what type of ethical principle in this situation?
    a. Paternalism
    b. Beneficence
    c. Autonomy
    d. Non-maleficence

2) The patient is clearly lucid and alert. What principle states that she has the right to make her own decision regarding medical treatment?
    a. paternalism
    b. Autonomy
    c. Beneficence
    d. Non-Maleficence

3) The patient decompensates. She is stabilized but not at this point alert. She is stuporous. The physician orders a G tube to be placed. What principle is applied here?
    a. paternalism
    b. Autonomy
    c. Beneficence
    d. Non-Maleficence

45-year-old female has a seizure at her house. She is found on MRI brain to have a large tumor. She undergoes emergency craniotomy for tumor resection. Pathology returns as small cell lung cancer. Her sister does not want the physicians to tell her about the diagnosis.

4) What principle does this violate?
    a. Full disclosure
    b. Principle of Habius Corpus
    c. Principle of Beneficence
    d. Principle of autonomy

5) The physician feel that to tell the patient her diagnosis a few hours after surgery would impair her healing and worsen her depression symptoms. He vows to tell her at a later point in care. What principle is this called?
    a. Autonomy
    b. Full disclosure
    c. Therapeutic privilege
    d. non-Maleficence

6) 6 months later this patient has a cardiac arrest. What documents must be identified with regard to treatment of this patient while having an arrest?
   a. Statement of living
   b. Birth records
   c. Living will or advance directive

7) An advance directive is found that does not stipulate a durable power of attorney. Who should be contacted next regarding future decision?
   a. Spouse
   b. Son
   c. Cousin
   d. Close uncle

34-year-old female is scheduled for elective appendectomy. Her general surgeon explains the risks of infection, bleeding, perforating bowel, and possible need for further surgeries. All of her questions are answered in detail.

8) The surgeon's explanation to the patient is known as what process?
   a. Complete consent
   b. Informative consent
   c. Informed consent
   d. Full consent

9) The patient decides that same day that she does not want to go through with the surgery at this point. The physician speaks with her regarding this but she remains steadfast. What should be done next?
   a. Respect her right to change her mind despite having signed a consent
   b. Force her to do surgery because that's what she needs
   c. Continue to speak with her until she changes her mind

10) Her son who is 16 has just married. He presents to the ED with acute appendicitis. He requires surgery. Who can sign the consent for surgery?
    a. The minor who is emancipated
    b. His mom only
    c. His father only
    d. His wife only

11) An 87-year-old male who lives in a nursing home is found to have new bruises each time his family comes to see him. The physician in the nursing home notes that his youngest son has a violent temper and frequently pushes his father around. What is the next step the physician should make?
    a. Do nothing
    b. Tell the head of the nursing home

c. Speak to the son
d. Report son to social services for elder abuse

12) A 3-year-old child is brought in to the ED by mother and father with a skull fracture and laceration on scalp. Upon further exam the child has multiple old fractures of femur and hemorrhages in the eyes. What is next treatment plan that the physician must make?
   a. Do nothing
   b. Report to child protective services and do Non accidental trauma workup
   c. Tell the family not to beat their child

13) A 65-year-old male patient who is hospitalized with pancreatic cancer tells the physician that he very angry with his cousin. He states his cousin has verbally abused him even while being in the hospital. He states that when he goes home he will confront his cousin in an aggressive manner. What is the responsibility of the physician at this time?
   a. Notify the third party
   b. Do nothing
   c. Call the cops immediately

14) You notice that your colleague in residency has been showing up to work with alcohol on his breath and has been acting very strange since he broke up with his girlfriend 5 months ago. His drinking has impaired his abilities as a surgeon as well and he has been neglecting much of his responsibilities at work. What should be done next?
   a. Report to the nursing director
   b. Call the cops
   c. Report to the next highest authority figure—In this case report to the chief resident who will report to the chairman

15) You are on a rotation as a resident with an attending physician who calls you at home stating that he has tooth pain that has been very severe and he requests that you write him a prescription for a narcotic. He states that nothing else helps this pain and that he is planning on seeing the dentist soon. What should you say?
   a. "I'm sorry, I cannot write prescriptions unless I have seen you and examined you myself. You may go to the ED for emergency care"
   b. give him the medication he requests
   c. give him only a few tablets until he can see his PCP
   d. give him Tylenol only

16) A 56-year-old female who has suffered severe neurologic injury after an aneurysm rupture is now comatose. Her family wants everything done at this point. They want aggressive measures despite the fact that her physicians expect no neurologic activity. What is the responsibility of the physicians to the family in this particular situation?
   a. No treatment required if treatment has no physiological basis but speak with family regarding comfort measures
   b. Take to the OR

c. Immediately remove Endotracheal tube

17) What principle is this called?
    a. Principle of Futility
    b. Principle of full disclosure
    c. Principle of uselessness

18) What committee may be consulted to help with this decision making issue?
    a. Court
    b. Operations committee
    c. Ethics Committee
    d. Graduate medical Education Committee

19) A surgeon accidentally operates on the wrong side of the abdomen during surgery. This has left the patient unable to tolerate feed by mouth. What is the responsibility of the physician post-operatively when speaking with the family?
    a. Immediately tell the patient that an error was made with a explanation and apology
    b. Don't tell the patient about it
    c. Tell the patient most of what happened
    d. Tell the patient that this mistake happens very commonly

20) An 87-year-old female has suffered a massive intracerebral hemorrhage. She is severely disabled and is hemodynamically unstable. She still opens her eyes to deep pain and has occasional non-purposeful movements of her arms and legs. Physicians expect this to be a non-survivable injury. She is placed on a morphine drip to diminish pain. What principle allows morphine to be used to not only relieve pain but to also hasten death?
    a. Principle of dual pain control
    b. Principle of Utility
    c. Principle of Hastening Death
    d. Double Effect Principle

21) 45-year-old female is s/p motor vehicle collision. She is neurologically devastated and in a coma. Her family wants to withdraw care in accordance with physician recommendations. What type of practice involves withdrawing of care that is legal in every state?
    a. Passive euthanasia
    b. Active Euthanasia
    c. Deactive Euthanasia

Answers: Ethics

1) Paternalism – wanting to, as physicians, make all of the decisions for our patients that may conflict with autonomy
2) Autonomy- patients are independent and may make their own decisions with regard to medical treatment
3) Beneficence- represents our obligation to help patients and act in their best interests
4) Full disclosure-Patients have a right to know what their diagnosis is; physicians cannot withhold information about a patient's diagnosis even if a family member does not want the patient to know. Always first ask the patient how much they want to know about the disease before telling them the diagnosis
5) Therapeutic Privilege- refers to the instance where the physician can withhold the patient's diagnosis if they believe that the information may harm the patient and cause their health to deteriorate further
6) Living will or advance directive- Living Will – a legal document, authored by a patient, that defines what medical procedures they would like to have done to them in case they are unable to make decisions on their own behalf (i.e., do not resuscitate, do not intubate). An Advanced Directive is like a living will, although in addition, it stipulates a durable power of attorney, who will be the patient's health proxy and thus make decisions on the patient's behalf if the patient is unable to or no longer capable of doing so. Those decisions will be made in light of thinking of how the patient would have handled the situation and with the patient's ethical principles in mind. Holds more precedence than a living will. Physicians must honor first what the patients want themselves as written in their advanced directive, regardless of what their health proxy may state. The durable power of attorney cannot make decisions against what is already written in the advanced directive, but can only add their decision if the decision is not clear or if another situation develops unpredicted. This of course keeping in mind what the patient would have wanted consistent with their own views. If no durable power of attorney is present or stated, the next of kin takes over the decision-making process, in this order: spouse, adult children, parents, siblings (oldest to youngest), friends, physician.
7) Spouse-see above
8) Informed consent

Involves describing the risks, benefits, alternatives, and consequences of getting or not getting a particular treatment or procedure, and requires their signature on a legal document
-Is not necessary in emergency treatment (i.e., it is implied) for minors (i.e., those under 18) or those who are altered and unable to make decisions for themselves
-The patient may decide that they no longer want treatment at any time after they sign the informed consent
-Informed consent is also important for entrance into a research study
      -Patients may leave the research study at any time
-Children require consent from adults
      -Excluding minors who are emancipated (i.e., married or in the armed forces) and those who seek treatment for sexually transmitted diseases, pregnancy, contraception, or drug-related issues
-If adult decides to not give a treatment to their child that physicians assume to be in the child's best interest, physicians can obtain a court order to treat the child

9) Respect her right to change her mind despite having signed a consent
10) The minor who is emancipated
11) Report son to social services for elder abuse
    Report to social services

-The patient is a harm to themselves (i.e., suicidal ideation, intent, or plan) or is in grave disability (i.e., unable to take care of themselves)

12) Report to child protective services and do Non accidental trauma workup
Child abuse (including rape)
-Report to child protective services immediately
-Physicians are not required to disclose to the parents that they have been reported for child abuse

13) Notify the third party--Tarasoff decisions
-If the patient threatens to harm a third party, physicians are obligated to attempt to notify the third party involved and to notify police to protect the third party

14) Report to the next highest authority figure—In this case report to the chief resident who will report to the chairman

15) "I'm sorry, I cannot write prescriptions unless I have seen you and examined you myself. You may go to the ED for emergency care"

-Avoid prescribing medications to patients unless you have examined the patient thoroughly and documented a written note

16) No treatment required if treatment has no physiological basis, if treatment has already been tried once and was unsuccessful and if maximal treatment has already been offered. Thus, no further treatment needed.

Principle of futility states that physicians are not required to treat if:
-Treatment has no physiological basis
-Treatment has already been tried once and was unsuccessful
-Maximal treatment has already been offered

17) Principle of Futility-see above

18) Ethics Committee—Every hospital usually has one.

19) Immediately tell the patient that an error was made with a explanation and apology

20) Double Effect Principle

physicians can prescribe medication that relieves pain even though it at the same time may hasten death (e.g., giving morphine to a patient with disseminated cancer during their final hours)

21) Passive euthanasia

– involves withholding care
-Legal in every state

Neurology

24-year-old female is cold. She wonders what nucleus in her brain controls heat production.

1) What region of the brain is involved with heat production?
    a. Posterior nucleus of Hypothalamus
    b. Medial hypothalamus
    c. Lateral hypothalamus
    d. Anterior hypothalamus

2) What region of the brain is involved with parasympathetic control to allow the body to cool down by sweating?
    a. Lateral hypothalamus
    b. Posterior nucleus of Hypothalamus
    c. Anterior nucleus of hypothalamus
    d. Medial hypothalamus

3) Her friend asks her why she is eating so much. She responds that the nucleus that controls eating too much is the:
    a. Lateral nucleus of hypothalamus
    b. Medial hypothalamus
    c. Posterior nucleus of Hypothalamus
    d. Anterior nucleus of hypothalamus

4) Her friend then asks here what nucleus is involved with limiting food intake?
    a. Medial nucleus of Hypothalamus
    b. Posterior nucleus
    c. Lateral nucleus
    d. Anterior nucleus

5) Destruction of what nucleus is involved with causing severe anger?
    a. Septal nucleus
    b. Nucleus solitarius
    c. Midbrain nucleus
    d. Hypothalamus

6) She has been having pain along the right side of her face since she was a child. She goes to the doctor and asks him what pathway is involved with facial sensation and through what thalamic nucleus does it travel?
    a. Trigeminal nerve, VPM
    b. Facial nerve, VPM
    c. Trigeminal nerve, VL
    d. Facial nerve, VPL

7) What hypothalamic nucleus controls circadian rhythms?
    a. Lateral nucleus
    b. Anterolateral nucleus
    c. Uncus
    d. Suprachiasmatic nucleus

A 24-year-old female is found to have significant placidity throughout her life, with interest in placing toys in her mouth with an increased sexual behavior than usual.

8) What syndrome is this?
    a. Prader willi syndrome
    b. Angelman Syndrome
    c. Kluver Bucy Syndrome
    d. Freudian Syndrome

9) This syndrome is caused by damage to what structure?
    a. Bilateral amygdala
    b. Posterior parietal lobe
    c. Dominant inferior frontal lobe
    d. Wernicke's Area

10) The amygdala is part of what lobe of the brain?
    a. Temporal—some classify in the limbic lobe
    b. frontal
    c. parietal
    d. cerebellum

11) What is the next branch off the external carotid artery after facial artery?
    a. Superficial temporal artery
    b. Occipital artery
    c. Lingual artery
    d. Internal maxillary artery

12) How many branches does the internal carotid artery have in the neck?
    a. None
    b. one
    c. two
    d. three

13) When performing a Lumbar puncture, the iliac crests are classically at what level?
    a. L4-5 Interspace
    b. L5-S1 Interspace
    c. L2-L3 Interspace
    d. L3-L4 Interspace

14) When performing a lumbar puncture, a "Pop" is felt when the needle penetrates what spaces?
    a. Dura
    b. pia
    c. spinal cord
    d. Epidural space

A 24-year-old female has pain that she describes as sharp and stabbing in nature in her left arm going down the lateral arm and forearm to the first and second digits. MRI of her cervical spine demonstrates degenerative changes of her spine.

15) What dermatome is associated with her symptoms?
    a. C6
    b. C7
    c. C8
    d. T1

16) At what level is the likely location for a cervical disc herniation?
    a. C5-6
    b. C2-3
    c. C4-5
    d. C8-T1

17) What nerve exits through the C4-5 foramen?
    a. C4
    b. C3
    c. C5
    d. C6

18) What nerve exits at the C7-T1 foramen?
    a. C8
    b. T2
    c. T1
    d. C7

19) Through what foramen does the C1 nerve root exit?
    a. There is no C1 nerve root
    b. C2-3
    c. C3-4
    d. Occipital-Axial

20) The C8 nerve root causes pain at what aspect of the hand?
    a. Medial 4th and 5th digits
    b. Lateral hand

        c. Back of the hand
        d. Thumb only

21) What electrical test can be used to diagnose a radiculopathy?
    a. EMG/NCS
    b. Tinel Test
    c. Phalens Test

22) The above patient has not had any treatment for her condition. What is the most appropriate initial treatment?
    a. Conservative therapy with PT/OT for 6 weeks
    b. Operate immediately
    c. Minimally invasive approach to remove the disc fragment
    d. Large doses of MS Contin to relieve her pain

23) PT does not help the patient. She has tried anti-inflammatory medication as well that have not worked. This has truly affected her quality of life. What is the next best treatment for this young lady with disc herniation?
    a. Continue further conservative treatment
    b. Operative management for discectomy with possible fusion
    c. Large doses of MS Contin to relieve her pain
    d. Nothing can be offered to relieve her pain

A 56-year-old male is an active runner but is now limited by pain in his right leg that is sharp in nature going down the posterior-lateral thigh, anterior shin, and ending in his big toe. The soles or lateral aspects of his foot are unaffected. He has some weakness flexing his big toe. MRI lumbar spine shows a disc herniation.

24) What is the likely location of this disc herniation if the disc was herniating far laterally?
    a. L5-S1
    b. S1-S2
    c. L3-L4
    d. None of the above

25) What if the disc was herniating centrally?
    a. L1-L2
    b. S1-S2
    c. L4-5
    d. None of the above

26) The L4 nerve root exits through which foramen?
    a. L4-5 foramen
    b. L3-4 foramen

    c. L1-2 Foramen
    d. None of the above

27) The L4 nerve root exits under what pedicle?
    a. L3 pedicle
    b. L2 pedicle
    c. L5 pedicle
    d. L4 pedicle

28) What is the most appropriate initial treatment for the above patient?
    a. Nothing can be done because it is too late
    b. Minimally invasive approach to remove disc
    c. OR immediately
    d. Conservative therapy with PT and anti-inflammatory medications

29) Baclofen may also be used to treat his pain. What is the main action of baclofen?
    a. Muscle relaxant
    b. Nerve pain reliever
    c. Opiate to numb the pain
    d. Strengthen the muscles

30) Baclofen works on what type of receptors?
    a. Ach
    b. GABA B receptor agonist
    c. GABA A receptor agonist
    d. GABA B receptor antagonist

31) Neurontin (Gabapentin) was prescribed for this patient. What is the goal of this medication?
    a. Decrease nerve pain
    b. Relax muscles
    c. Strengthen the muscles of the lower back
    d. Stop the focal back pain

32) The above patient was found to have a very hypoactive right Achilles reflex. What nerve does this localize to?
    a. Right L3
    b. Right L4
    c. Right S1
    d. Left S1

33) The L5 nerve root is involved in what lower extremity reflex?
    a. Achilles
    b. L5 is involved in no reflex

c. Patellar
  d. brachioradialis

34) Triceps reflex is associated with what nerve?
   a. C7
   b. C6
   c. C5
   d. C8

35) A positive Hoffman sign indicates what pathology?
   a. Cervical Myelopathy
   b. radiculopathy
   c. muscle cramps
   d. herniated lumbar disc

36) An upgoing babinski reflex indicates what pathology in a 45 year old male?
   a. Lower motor neuron sign
   b. Upper motor neuron sign on ipsilateral side
   c. It does not mean anything
   d. Upper motor neuron pathology on contralateral side brain or ipsilateral spinal cord

A 34-year-old male is stabbed along his right shoulder. He has severe right arm weakness.

37) EMG/NCS done states that his axillary nerve has been damaged. What cord of the brachial plexus does this come off?
   a. Posterior
   b. Anterior
   c. medial
   d. Lateral

38) The axillary nerve innervates what muscles?
   a. Teres major and Deltoid
   b. Deltoid only
   c. Teres minor and Teres Major
   d. Deltoid and teres minor

39) The musculocutaneus nerve innervates what muscles?
   a. Biceps and Brachialis
   b. Brachialis and Teres Major
   c. Corachobrachialis and brachialis
   d. Biceps, coracobrachialis, and brachialis

A 56-year-old female typist presents with pain and tingling over the palm of her right hand affecting the first four digits. The pain wakes her up at night.

40) What physical exam tests can be done to confirm the diagnosis?
    a. Cubital tunnel sign
    b. Only Tinel sign
    c. Phalen sign and Tinel sign
    d. Adson's Test
    e. Patrick's Test

41) What electrical study can be ordered to confirm the diagnosis?
    a. EMG/NCS
    b. Tinel's Test
    c. Phalen's Test
    d. Spurling Sign

42) The NCS is positive for right hand carpal tunnel syndrome. After a trial of splints and medication, her pain has continued. What is next treatment option for this typist?
    a. Nothing can be offered
    b. Continue with splints
    c. OR for carpal tunnel release
    d. OR for bilateral carpal tunnel release

43) In a carpal tunnel release, what ligament is incised?
    a. Transverse carpal ligament
    b. Cubital tunnel ligament
    c. Frohse ligament
    d. None of the above

A 45-year-old male athlete presents with left hand pain affecting his last two digits. He has developed a claw deformity of the last two fingers of his hand.

44) What is the likely diagnosis?
    a. Left ulnar neuropathy
    b. Left carpal tunnel syndrome
    c. Right ulnar neuropathy
    d. Entrapped suprascapular nerve

45) How might one localize the site of ulnar nerve entrapment?
    a. MRI
    b. EMG/NCS
    c. CT
    d. None of the above

46) What physical exam can be done to diagnose an ulnar neuropathy at the elbow?

a. Palpate and tapping over medial epicondyle can reproduce exact symptoms
   b. Turning head to ipsilateral side to reproduce Symptoms
   c. Tap over transverse carpal ligament
   d. Flexing hip and reproducing the Symptoms

47) Ulnar neuropathy at the elbow is most commonly caused by entrapment at what canal?
   a. Cubital tunnel
   b. Ligament of Frohse
   c. Guyon's canal
   d. None of the above

48) What nerve injury can cause a wrist drop?
   a. Musculocutaneous nerve
   b. Radial nerve-C7
   c. Median nerve-C6
   d. None of the above

49) What cord of the brachial plexus does the radial nerve come off of?
   a. Posterior
   b. Anterior
   c. Medial
   d. Lateral

50) What other pathology can lead to a wrist drop?
   a. Lead poisoning
   b. Manganese poisoning
   c. Copper poisoning
   d. Iron deficiency

51) Ulnar neuropathy can lead to atrophy of what muscles in the hand?
   a. Thenar muscles
   b. Brachioradialis
   c. Hypothenar muscles
   d. None of the above

52) Conservative therapy has not helped the above patient with ulnar neuropathy. What is next treatment approach?
   a. Ulnar nerve decompression at cubital tunnel
   b. Nothing to do
   c. Continue splints
   d. Carpal tunnel release

53) EMG/NCS shows ulnar neuropathy at wrist. What canal is this?
   a. Guyon's Canal

b. Cubital tunnel
c. Carpal tunnel
d. None of the above

54) Ulnar neuropathy at Guyon's Canal has not responded to conservative therapy. What is next best treatment option?
   a. Decompression at elbow
   b. Nothing to do
   c. Continue pain medication for the pain
   d. Decompression at wrist-medial to carpal tunnel site

Carpal tunnel syndrome is a common medical condition that affects many people in the world.

55) Which of the following can cause this syndrome?
   a. Pregnancy, acromegaly, hypothyroidism
   b. Hyperparathyroidism and Acromegaly
   c. Acromegaly and Hyperthyroidism
   d. Hyperthyroidism only

A 56-year-old female is s/p breast mastectomy for breast cancer. She has severe pain on the right side of her back and has noticed that a bone on her back is pushing out when she pushes against a wall.

56) What is the likely diagnosis?
   a. Hinged clavicle
   b. Bow hunter's Syndrome
   c. Winged scapula
   d. None of the above

57) What nerve was damaged during her mastectomy?
   a. Long thoracic nerve
   b. Short thoracic nerve
   c. Dorsal scapular nerve
   d. Nerve to subclavius

58) What muscle does the long thoracic nerve innervate?
   a. Subclavius
   b. Serratus posterior
   c. Serratus anterior
   d. Sternocleidomastoid

A 45-year-old male has a large brain tumor in the pituitary gland that has caused right eye blindness.

59) What cranial nerve is being impinged on by tumor?
   a. CN 3
   b. CN 2 optic nerve
   c. CN 4
   d. CN 5

60) What if the patients tumor was in the medial temporal lobe and has caused his right eye to become ptotic with a dilated and unresponsive pupil, with loss of extraocular movement of the right eye. What nerve is being impinged?
   a. CN 3
   b. CN 2
   c. CN 4
   d. CN 6

61) A pituitary tumor pushing on the optic chiasm classically causes what visual pathology?
   a. Right homonymous hemianopsia
   b. Left homonymous hemianopsia
   c. Hemianopsia with macular sparing
   d. Bitemporal hemianopsia

62) A tumor pushing on the right optic tract causes what visual pathology?
   a. Left homonymous hemianopsia
   b. Bitemporal hemianopsia
   c. Right homonymous hemianopsia
   d. Hemianopsia with macular sparing

63) A lesion in Meyers loop in right inferior temporal lobe causes what visual pathology?
   a. Left pie in sky appearance-superior quadrantanopsia
   b. Right pie in sky appearance-superior quadrantanopsia
   c. Right homonymous hemianopsia
   d. Hemianopsia with macular sparing

64) A lesion in right occipital lobe causes what visual pathology?
   a. Left pie in sky appearance-superior quadrantanopsia
   b. Right homonymous hemianopsia
   c. Right pie in sky appearance-superior quadrantanopsia
   d. Left homo hemianopsia with macular sparing

65) The abducens nerve causes movement of the eye in what direction?
   a. Laterally
   b. Posteriorly
   c. Medially
   d. Superiorly

A 56-year-old male suffered a massive traumatic brain injury with multiple fractures through the skull base. His tongue has now deviated to the right side.

66) What side is the fracture through the hypoglossal canal?
    a. Left side
    b. Fracture cannot cause this pathology
    c. Right side

A 24-year-old male presents with personality changes and choreiform movements in his arms and legs. His mother has this same pathology.

67) What is the likely diagnosis?
    a. Huntington disease
    b. Picks Disease
    c. Niemann-Pick Disease
    d. Tay-Sach's Disease

68) What genetic pattern does this follow?
    a. Aut Recessive
    b. X linked Recessive
    c. Aut Dominant
    d. Mitochondrial Inheritance

69) What chromosome is this gene on?
    a. Chrom 4
    b. Chrom 5
    c. Chrom 6
    d. Chrom 18

70) A 23-year-old male presents with seizures. He has a birthmark on his lower back and areas of hypopigmented regions on his skin. What is the likely diagnosis?
    a. Niemann-Pick's Disease
    b. Tay-Sach's Disease
    c. Von Hippel Lindau Disease
    d. Tuberous Sclerosis

71) What is the genetic pattern?
    a. Aut Dominant, chrom 16
    b. Aut Recessive, Chrom 16
    c. X linked Dominant
    d. Mitochondrial Inheritance

72) What study should be done next to look for other abnormalities associated with the disease?
    a. MRI-Look for renal cysts
    b. PET scan-Look for "Hot" Tumors

c. CT Head/Chest /abdomen/pelvis—angiomyolipomas, cardiac rhabdomyomas, astrocytomas
   d. No further imaging needed

73) CT head classically shows what pathology?
   a. Calcified tubers in periventricular regions
   b. Multiple scattered hemorrhages throughout the brain
   c. Multiple skull fractures in different stages of healing
   d. none of the above

74) Skull X-ray classically shows what pathology?
   a. No abnormality can be seen on X-ray in these patients
   b. Tram Track calcification
   c. Multiple nevi calcified
   d. None of the above

Answers: Neurology

1) Posterior nucleus of Hypothalamus
2) Anterior nucleus of Hypothalamus
3) Lateral nucleus of hypothalamus
4) Medial nucleus of Hypothalamus
5) Septal nucleus
6) Trigeminal Nerve, VPM
7) Suprachiasmatic nucleus
8) Kluver-Bucy Syndrome:
   -Caused by bilateral destruction of the amygdala
   -Hyperorality
   -Hypersexuality
   -Docile

9) Bilateral amygdala-see above
10) Temporal-some classify in limbic lobe
11) Occipital artery

Branches of External Carotid Artery
-Superior Thyroid Artery
-Ascending Pharyngeal Artery
-Lingual Artery
-Facial Artery
-Occipital Artery
-Posterior Auricular Artery
-Superficial Temporal Artery
-Internal Maxillary Artery

Branches of Internal Carotid Artery
-Has no branches in the neck
-Has many branches in the brain

12) None
13) L4-5 Interspace
14) As it goes through Dura

   Skin → Ligaments (i.e., supraspinous, interspinous, ligamentum flavum) → Epidural space → Dura (i.e., hear classic "pop") → Subdural space → Arachnoid → Subarachnoid space (i.e., location of CSF)
   -The pia is not pierced during an LP

15) C6
16) C5-6
17) C5
18) C8
19) No C1 nerve root
20) Medial 4th and 5th digits
21) EMG/NCS
22) Conservative therapy with PT/OT for 6 weeks
23) Operative management for discectomy with possible fusion

24) L5-S1
25) L4-L5
26) L4-5 foramen
27) L4 Pedicle
28) Conservative therapy with PT and anti inflammatory medications
29) Muscle Relaxant
30) GABA B receptor agonist
31) Decrease nerve pain
32) Right S1
33) L5 has no reflex
34) C7
35) Cervical Myelopathy
36) Upper motor neuron pathology on contralateral side brain or ipsilateral spinal cord
37) Posterior
38) Deltoid and teres minor
39) Biceps, coracobrachialis, and brachialis
40) Phalen sign and Tinel sign
41) EMG/NCS
42) OR for carpal tunnel release
43) Transverse Carpal Ligament
44) Left ulnar neuropathy
45) EMG/NCS
46) Palpate and tapping over medial epicondyle can reproduce exact symptoms
47) Cubital Tunnel
48) Radial nerve-C7
49) Posterior
50) Lead Poisoning
51) Hypothenar muscles
52) Ulnar nerve decompression at cubital tunnel
53) Guyons Canal
54) Decompression at wrist-medial to carpal tunnel site
55) Pregnancy, acromegaly, hypothyroidism
56) Winged scapula

Damage to long thoracic nerve that innervates the serratus anterior muscle
-Commonly injured after breast mastectomies or trauma
-Pushing against a wall reveals that the scapula protrudes outward unilaterally

57) Long thoracic nerve—see above
58) Serratus Anterior—see above
59) CN 2 optic nerve
60) CN 3
61) Bitemporal hemianopsia
62) Left homonymous hemianopsia
63) Left pie in sky appearance-superior quadrantanopsia
64) Left homo hemianopsia with macular sparing
65) Laterally
66) Right side
67) Huntington Disease
68) Aut Dominant

69) Chrom 4
70) Tuberous Sclerosis
71) Aut Dominant, chrom 16
72) CT Head/Chest /abdomen/pelvis—angiomyolipomas, cardiac rhabdomyomas, astrocytomas
73) Calcified tubers in periventricular regions
74) Tram Track calcification

Psychiatry

24-year-old male has been training for the Olympics while studying for midterms in college. He recently was forced to miss a meet due to a final exam. He has been angered by that and this has made him more aggressive in is training regimen.

1) What defense mechanism is he exhibiting?
    a. Sublimation
    b. Humor
    c. Suppression
    d. Regression

2) What if he began joking about his missing the Olympics because of other exams. What Defense mechanism is this?
    a. Regression
    b. Suppression
    c. Sublimation
    d. Humor

3) Is Humor a mature or immature defense mechanism?
    a. Regression
    b. Suppression
    c. Humor
    d. Sublimation

4) He decides to stop thinking about the Olympics as he has been extremely nervous of late. What defense mechanism is this?
    a. Suppression
    b. Humor
    c. Sublimation
    d. Regression

5) He was able to participate in the Olympics but lost every event. He was very depressed for quite a while afterwards but has seemingly recovered. His friends ask him about the Olympics but he keeps changing the topic to avoid any discussion of it. He occasionally states that he did not really lose in the Olympics. What defense mechanism is this?
    a. Denial
    b. Suppression
    c. Regression
    d. Anger

6) His mom attempts to talk to him but he lashes out at her. He starts screaming and acting out in ways that he did when he was a child. What defense mechanism is this?
    a. Rationalization
    b. Suppression
    c. Intellectualization
    d. Regression

7) Is regression a mature or immature defense mechanism?
    a. Immature
    b. Mature
    c. None of the above

8) He comes home one day and decides to yell at his dog for no apparent reason. He even decides not to walk his dog, a route he would do every night. What defense mechanism is this?
    a. Repression
    b. Anger
    c. Denial
    d. Displacement

9) What defense mechanism is associated with a person taking an opposite view to avoid the anxiety associated with a prior view point?
    a. Reaction formation
    b. Denial
    c. Anger
    d. Sublimation

A 45-year-old recently divorced female states that she hates all men.
10) What defense mechanism is this?
    a. Splitting
    b. Reaction formation
    c. Suppression
    d. None of the above

11) When someone focuses on the minute details of a topic that causes further anxiety, this is called what type of defense mechanism?
    a. Intellectualization
    b. De intellectualization
    c. Rationalization
    d. None of the above

A 56-year-old obese male snores at night. He is noticed to awaken multiple times during the night. He has excessive daytime sleepiness.

12) What is the most likely diagnosis?
    a. Obstructive sleep apnea
    b. Night terrors
    c. REM Sleep
    d. None of the above

13) What is the method for diagnosis?
    a. polysomnography to measure the number of apnea/hypopnea episodes
    b. physical exam
    c. EMG/NCS
    d. History only

14) What is best treatment for these patients?
    a. Weight loss and CPAP at night
    b. Exercise only
    c. Steroid inhalers at night
    d. none of the above

A 34-year-old male has episodes of falling immediately to the ground after emotional states. He is noted to have sudden episodes of falling asleep as well.

15) What is diagnosis?
    a. Narcolepsy
    b. Night terror
    c. Sleep apnea
    d. Insomnia

16) He is noted to have hallucinations when falling asleep. What are these called?
    a. Dreaming hallucinations
    b. Hypnopompic hallucinations
    c. Hypnagogic hallucinations
    d. None of the above

17) What is best method for diagnosis?
    a. CT head
    b. CT Chest
    c. MRI Brain
    d. History and polysomnography

18) What is treatment of choice?
    a. Ritalin, scheduled naps
    b. Exercise
    c. caffeine
    d. No treatment available

19) Ritalin is what type of medication?
    a. Stimulant (amphetamine)
    b. opiate
    c. anti-inflammatory
    d. steroid

A 23-year-old male often times is noted by his wife to awaken abruptly screaming with sweating episodes. He does not recall what he was dreaming about.

20) What is the diagnosis?
    a. Night terrors
    b. nightmares
    c. REM sleep
    d. None of the above

21) Does this occur during REM or NREM sleep?
    a. NREM-stage 3 or 4
    b. REM
    c. NREM 1
    d. NREM 2

A 34-year-old male obtains sexual arousal from being hurt by sexual partner.
22) What paraphilia is this called?
    a. fetishism
    b. voyeurism
    c. pedophilia
    d. Masochism

23) What if he obtained arousal from nonliving object?
    a. masochism
    b. voyeurism
    c. Fetishism
    d. sadism

A 43-year-old male is being treated for depression with SSRI. He has noted that he has lost his erectile ability over the last few months. He is seeking advice from his physician.

24) What is the first treatment for this patient?
    a. Stop the SSRI
    b. Send him to a gynecologist
    c. Reduce stress in his life with exercise and counseling
    d. There is no treatment for this

25) If his erectile dysfunction has not been treated by the above methods, what other medical treatment can be used?
    a. oxycodone
    b. xanax
    c. Viagra-Sildenafil
    d. None of the above

26) How does Viagra work?
    a. Phosphodiesterase 5 inhibitor-increasing level of CGMP
    b. ACH receptor agonist
    c. NE releaser
    d. Inhibits release of NE from pre synaptic terminal

75-year-old male with 4 weeks history of hypersomnia, guilt, psychomotor depression with occasional agitation, hyperphagia and weigh gain. He notes that he had one episode of suicidal ideation.

27) What is the diagnosis?
    a. Major Depressive Disorder
    b. Atypical depression
    c. Anxiety attacks
    d. None of the above

28) What is the treatment of choice?
    a. SSRI
    b. oxycodone
    c. xanax
    d. requip

29) What if the patient only had symptoms of weight gain, leaden paralysis and rejection sensitivity. What is the diagnosis?
    a. Major Depression
    b. Panic attacks
    c. Seasonal depression
    d. Atypical Depression

30) What is treatment of choice for atypical depression?
    a. MAO I
    b. SSRI
    c. codeine
    d. none of the above

A patient recently being treated via medication for schizophrenia develops acute onset toxic appearance with fevers and muscle rigidity and seizures. He has an elevated wbc count and CK level.

31) What is the most likely diagnosis?
    a. Neuroleptic malignant syndrome
    b. Serotonin syndrome
    c. Malignant hyperthermia
    d. None of the above

32) What is the next treatment choice?
   a. No treatment exists
   b. Give narcan
   c. Give flumazenil
   d. Stop all antipsychotic medication, resuscitate in ICU, dantrolene

A 24-year-old female is noted to have high fevers to 41 C intra-operatively during appendectomy immediately after being given paralytic agents. She has a family history of anesthesia sensitivity.

33) What is the diagnosis?
   a. Malignant Hyperthermia
   b. Serotonin syndrome
   c. Neuroleptic malignant syndrome
   d. None of the above

34) What are the treatments of choice?
   a. Give fentanyl
   b. Give fumazenil
   c. Dantrolene, cooling agents
   d. No treatment exists

35) Dantrolene works by:
   a. Decreasing release of calcium from sarcoplasmic reticulum and binding to ryanodine receptor
   b. Ach released from terminal
   c. Inhibits NE release from terminal
   d. Releases Dopamine from nerve terminals

Answers: Psychiatry

1) Sublimation

Sublimation – shifting anger towards goal-directed activities (i.e., using anger to exercise harder)
    -Mature defense mechanism
-Humor – using humor to lessen the anxiety (i.e., someone who makes a joke about their situation during a period of stress)
    -Mature defense mechanism
-Suppression – consciously, keeping anxiety-provoking thoughts out of conscious awareness (i.e., medical student who chooses to stop thinking about their USMLE Step 2 exam until the exam looms closer)
    -Mature defense mechanism
-Denial – consciously avoiding the awareness of an anxiety-provoking thought/event (i.e., common in AIDS and cancer patients once given diagnosis)
    -Immature defense mechanism
-Regression – acting out, behaving as though one was a child (i.e., children who wet the bed once hospitalized)
    -Immature defense mechanism
Repression – unconsciously, thoughts are relegated out of conscious awareness (i.e., child who is molested forgets event happened)
    -Immature defense mechanism
-Displacement – displacing anger towards another object or person less threatening (i.e., kicking the dog when coming home from work after a long day)
    -Immature defense mechanism
-Projection – projecting one's feelings to another (i.e., saying another person hates you when you hate them)
    -Immature defense mechanism
-Reaction formation – Taking the polar opposite view so as to not experience the anxiety associated with a certain point of view or action (i.e., a man who was convicted of domestic violence raises money for a women's shelter)
    -Immature defense mechanism
-Splitting – viewing a person as all good or all bad with no middle-ground (i.e., patient who likes a medical student, but then hates the entire nursing staff)
    -Immature defense mechanism
    -Commonly seen in borderline personality disorder
-Rationalization – using excuses to help explain away a certain anxiety-provoking event or thought usually to avoid self-blame (i.e., person who gets fired says the boss had a personal vendetta against them, instead of blaming the firing on their poor work ethic)
    -Immature defense mechanism
-Intellectualization – focusing on the minute details of subject matter that is anxiety-provoking when thought of as a whole (i.e., son who examines the pathophysiology of prostate cancer once his father his diagnosed with that disease)
    -Immature defense mechanism

2) Humor
3) Mature
4) Suppression
5) Denial
6) Regression
7) Immature

8) Displacement
9) Reaction formation
10) Splitting
11) Intellectualization
12) Obstructive Sleep Apnea

Obstructive Sleep Apnea
-Seen in obese patients
-History of daytime somnolence or non-restful sleep
-Snoring during the night; catching one's breath during sleep
    -Partner usually can verify snoring
-Notable physical findings:
    -Large uvulas/tongues
    -Small jaws
    -Thick necks
-Diagnose with polysomnography to measure the number of apnea/hypopnea episodes
-Treat with weight loss for mild to moderate cases
-Consider adding nasal CPAP during nighttime for moderate cases
-Treat definitively with uvulopharyngoplasty for severe cases (i.e., specifically those with heart failure)
-Complications include cor pulmonale (i.e., pulmonary hypertension secondary to hypoxic vasoconstriction with normal left heart function)
-Associated with Sudden Infant Death Syndrome

13) polysomnography to measure the number of apnea/hypopnea episodes
14) Weight loss and CPAP at night
15) Narcolepsy

Narcolepsy
-More common in younger patients
-Daytime somnolence
-Sleep findings:
    -Sleep attacks (i.e., sudden episodes of falling asleep at any time)
    -Cataplexy – sudden loss of muscle tone with laughing and extreme emotional state
    -Sleep paralysis upon awakening
    -Hallucinations
        -Hypnagogic hallucinations – hallucinations as patient is falling asleep
        -Hypnopompic hallucinations – hallucinations as patient is waking up
    -Decreased REM latency – enter REM sleep earlier in the sleep cycle (i.e., enter REM sleep immediately as one drifts asleep)
-Diagnose with polysomnography
-Treat with scheduled naps and stimulants (i.e., amphetamines like Ritalin)

16) Hypnagogic hallucinations
17) History and Polysomnography
18) Ritalin, scheduled naps
19) Stimulant(Amphetamine)
20) Night terrors

21) NREM-stage 3 and/or 4
22) Masochism

Paraphilias
-Preoccupation with sexual fantasies
-Usually seen in men
-Exhibitionism – arousal from exposing one's genitals to a stranger
-Sadism – arousal from inflicting suffering on sexual partner
-Masochism – arousal from one being hurt by sexual partner
-Frouterism – touching or rubbing one's genitals along non-consenting person (i.e., common in subways and elevators)
-Fetishism – use of nonliving objects for sexual arousal
-Voyeurism – arousal obtained from viewing person who is in sexual state
-Pedophilia – arousal from sexual activity with minor

Gender Identity Disorder
-Confusion about one's gender identification
    -Gender – psychological/societal
    -Sex – biological/chromosomal
-More common in men
-Often exhibit cross-dressing
-May take hormones to change gender
-May pursue reconstructive procedures to alter their gender
-Treatment is supportive psychotherapy and hormonal therapy

23) Fetishism
24) Stop the SSRI
25) Viagra-Sildenafil

Erectile Dysfunction
-Causes:
    -Medications – β blockers, antihypertensives, alcohol, selective serotonin reuptake inhibitors (SSRIs)
    -Psychological – examine relationship (i.e., cheating in relationship)
        -Look to see if patient has had erectile problems with previous partners
            -If so, not likely psychological
        -Examine if patient has erections in the morning upon awakening
            -If so, likely psychological
    -Other comorbidities – hypertension, diabetes mellitus, coronary artery disease, atherosclerosis
        -Examine to see if blood pressure and glucose well-controlled
    -Post-surgery – especially after prostate removal
-Treatment is dependent on cause
    -If medication-induced, stop the medication suspected, if possible
    -If psychological, consider counseling and psychiatric consultation
    -If uncontrolled hypertension or diabetes mellitus, control underlying disease
        -Consider Viagra (i.e., sildenafil)
            -A phosphodiesterase, PDE5, inhibitor, increasing levels of cGMP relaxing smooth muscle
        -Consider implants for severe cases

26) Phosphodiesterase 5 inhibitor-increasing level of CGMP
27) Major Depressive Disorder
28) SSRI
29) Atypical Depression

Atypical Depression
- Atypical symptoms of depression:
    - Weight gain
    - Hypersomnia
    - Leaden paralysis
    - Rejection sensitivity
- Treat with MAO Inhibitors

30) MAO I
31) Neuroleptic malignant syndrome

Neuroleptic Malignant Syndrome (NMS)
-Signs and symptoms include:
    -Toxic appearance
    -High spiking fevers
    -Muscular rigidity
    -Seizures
    -Altered mental status
-Leukocytosis
-Elevated creatine kinase (CK)
-Life-threatening
-Can occur immediately after a patient starts a new antipsychotic medication
-Treat in ICU setting with ABCs
-Stop all antipsychotic medications
-Give dantrolene which prevents calcium to be released from the sarcoplasmic reticulum, or give bromocriptine a dopamine agonist

32) Stop all antipsychotic medication, resuscitate in ICU, dantrolene
33) Malignant Hyperthermia
34) Dantrolene, cooling agents
35) Decreasing release of calcium from sarcoplasmic reticulum and binding to ryanodine receptor

Ophthalmology

78-year-old Caucasian female presents to the ED with loss of vision in right eye. She denies pain. She denies trauma. She is a smoker with hypertension. On exam, she has loss of central vision in her right eye.

1) What is the most likely diagnosis?
    a. Macular Degeneration
    b. Closed angle glaucoma
    c. Open angle glaucoma
    d. Clouding of the lens syndrome

2) What consultation should be placed?
    a. Respiratory
    b. Optometry
    c. Cardiology
    d. Ophthalmology

3) What does the ophthalmologist see on fundoscopic exam characteristic for macular degeneration?
    a. Retinal detachment
    b. Papilledema
    c. Lens clouding
    d. Drusen

4) What is best treatment at this time if patient has exudative macular disease?
    a. Eyedrops only
    b. Laser Photocoagulation
    c. steroids
    d. none of the above

45-year-old African American male notices acute onset of red and tender right eye. He has pain and blurry vision in that eye as well. On exam he has a dilated right pupil that is responsive to light.

5) What is the most likely diagnosis?
    a. Closed Angle Glaucoma
    b. Open angle glaucoma
    c. Temporal arteritis
    d. Retinal detachment

6) Ophthalmology sees the patient. They note increased pressure in the right eye. What is treatment of choice at this time?
    a. Acetazolamide (diamox) IV or eye drops
    b. IVF
    c. steroids

d. none of the above

7) How does diamox work?
   a. Blocking formation of Norepinephrine
   b. Carbonic anhydrase inhibitor and decreases production of bicarbonate
   c. Stimulating increased drainage from the eye
   d. None of the above

8) What other medication can be used?
   a. Beta Blockers
   b. Alpha blockers
   c. Anti-inflammatory meds
   d. Aspirin

9) What medication should never be used in patients with glaucoma?
   a. Anti cholinergics
   b. Beta blocker
   c. steroids
   d. none of the above

10) Definitive treatment for closed angle glaucoma is:
    a. enucleation
    b. eye drops
    c. Laser iridotomy
    d. None of the above

A 67 male with diabetes notices gradual onset of vision loss in both eyes without pain. He has an increase in intraocular pressure on both eyes on exam but not red or tender.

11) What is most likely diagnosis?
    a. Open angle (chronic) Glaucoma
    b. Closed angle glaucoma
    c. Retinal detachment
    d. Pituitary tumor

12) What is treatment of choice?
    a. Acetazolamide
    b. steroids
    c. timolol eye drops
    d. clonidine

45-year-old male with uncontrolled diabetes and hypertension presents to ED with sudden loss of vision in most of his right eye. He denies pain or trauma. He has a long history of smoking. Ophthalmologic exam finds a cherry red spot on macula and a pale fundus.

13) What is most likely diagnosis?
    a. Central retinal artery occlusion
    b. Tay Sachs disease
    c. Temporal Arteritis
    d. None of the above

14) This is identified within 4 hours of onset. What is treatment of choice?
    a. Thrombolysis with heparin or TPA
    b. Intra-arterial TPA
    c. OR for enucleation of eye
    d. Spray eye drops

15) Patient is then found to have tender spot on right side of scalp with dilated temporal vein. What is most likely diagnosis?
    a. Central retinal artery occlusion
    b. Traumatic injury
    c. Superficial temporal arteritis induced blindness
    d. Congenital defect to the eye

16) What is treatment of choice?
    a. Immediate IV prednisolone
    b. Aspirin
    c. Beta blockers
    d. opiates

67-year-old female noted on eye exam to have dilated tortuous veins and retinal hemorrhages with cotton wool spots. She presented with painless vision loss.

17) What is most likely diagnosis?
    a. Central retinal vein occlusion
    b. Temporal arteritis
    c. Wallenberg syndrome
    d. Traumatic disruption of the eye

18) What is treatment of choice?
    a. Enucleate the eye immediately
    b. Spray antibiotic eye drops
    c. Laser photocoagulation
    d. None of the above

24-year-old female prostitute presents with skin lesions. On exam of her pupils, she has right pupil that does not constrict to light but does accommodate.

19) What is her diagnosis?

a. Neurosyphilis
   b. Wernicke syndrome
   c. Korsakoff syndrome
   d. Whipple's Disease

20) What is the name given to the pathology that is noted to affect her pupil?
   a. Argyll-Robertson Pupil
   b. Marcus Gunn Pupil
   c. Roving pupil
   d. None of the above

Answers: Ophthalmology

1) Macular Degeneration

Age-Related Macular Degeneration (ARMD)
-Most common cause of blindness of those greater than 65
     -Second most common cause is diabetes
-Risk factors include:
     -Age
     -Caucasians
     -Female
     -Family history
     -Smoking
     -Hypertension
     -Obesity
-Painless loss of central vision
-Two types:
     -Atrophic macular degeneration – causes gradual vision loss
     -Exudative macular degeneration - more rapid and severe vision loss
-See drusen (i.e., yellow deposits) in the macula on ophthalmoscopic exam
   -Treatment is usually none but may involve laser photocoagulation to delay loss of vision in exudative macular disease

2) Ophthalmology
3) Drusen
4) Laser Photocoagulation
5) Closed angle Glaucoma

Closed-Angle Glaucoma (Acute Glaucoma)
-Acute onset
-Typical history is person sitting watching a movie in dark theater with sudden blurry vision that is usually unilateral
-Signs and symptoms include:
     -Tender, hard eye
     -Red eye
     -Dilated pupil that is semi-responsive to light
-Obtain stat ophthalmologic consult
-Conduct tonometry (i.e., measure pressure in the eye) for diagnosis
-Treat with IV acetazolamide or acetazolamide eye drops
     -Carbonic anhydrase inhibitors – decrease the production of bicarbonate
-Can also give β blocker eye drops (i.e., timolol)
-Consider cholinergic eye drops (i.e., pilocarpine)
     -Never give anticholinergics to patients with glaucoma
-Definitive treatment is laser iridotomy

Open-Angle Glaucoma (Chronic Glaucoma)
-More common than acute glaucoma
-Can be bilateral
-Occurs more gradually as vision is lost
-Painless

-Treatment is to decrease intraocular pressure with topical eye drops (i.e., acetazolamide, timolol, or pilocarpine)
      -Never give anticholinergics
-Definitive treatment is laser iridotomy

6) Acetazolamide(diamox) IV or eye drops
7) Carbonic anhydrase inhibitor and decreases production of bicarbonate
8) Beta Blockers
9) Anti cholinergics
10) Laser Iridotomy
11) Open angle (chronic) Glaucoma-see above
12) Acetazolamide
13) Central retinal artery occlusion

Central Retinal Artery Occlusion
-May be from emboli
-May be from atherosclerosis
-Risk factors include:
      -Diabetes
      -Coronary artery disease
      -Hypertension
      -Smoking
      -Temporal arteritis
-Sudden blurry vision or loss of vision in one eye
-Painless
-May respond poorly to light, but will constrict abruptly when light shined in other eye
-See cherry red spot on the macula and a pale fundus
-Treat with thrombolysis if within 8 hours since symptom onset
-Consider IV acetazolamide or timolol to decrease intraocular pressure
-Treat risk factors
-Prescribe aspirin daily
-Digital massage of closed eye to dislodge emboli to block smaller artery leading to smaller area of ischemia in the retina
-If suspect may be secondary to temporal arteritis, start IV prednisolone immediately and schedule temporal artery biopsy

Central Retinal Vein Occlusion
-Associated with diabetes and glaucoma
-Visual loss is painless and occurs gradually
-Ophthalmologic exam is critical
      -Dilated tortuous veins
      -Retinal hemorrhages
      -Cotton-wool spots
      -Macular edema
-Treat with laser photocoagulation
      -Especially for diabetic retinopathy with neovascularization and branch retinal vein occlusion
-Treatment is to reduce risk factors
-Prescribe aspirin daily

14) Thrombolysis with heparin or TPA
15) Superficial temporal arteritis induced blindness
16) Immediate IV prednisolone
17) Central retinal vein occlusion—see above
18) Laser photocoagulation
19) Neurosyphilis
20) Argyll-Robertson Pupil

Argyll-Robertson Pupil
-Seen almost exclusively in neurosyphillis
-Pupil does not constrict to light, but does accommodate
	-Like a prostitute who accommodates but does not react
-Treat neurosyphillis with ceftriaxone

Marcus-Gunn Pupil
-Afferent pupillary defect
-Seen in tumors of the optic nerve or a complication of optic neuritis seen in multiple sclerosis
-Diagnose with swinging flashlight test
	-When place light in right eye, see both eyes constrict
	-When swing light to the left eye, no constriction seen in either eye
	-Suggests afferent problem in the left eye
-Treat underlying condition

Cardiovascular

45-year-old male presents with shortness of breath. EKG is shown below.

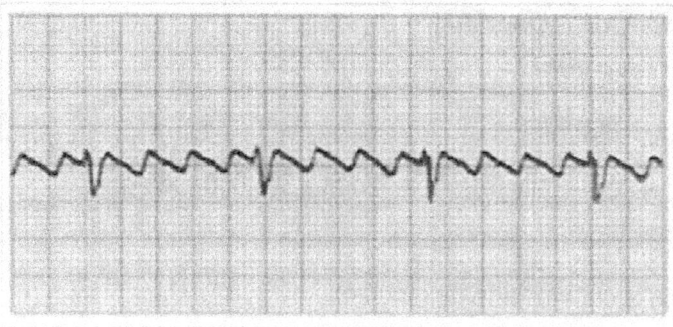

1. What is the most likely diagnosis?
    a. Atrial Flutter
    b. Atrial fib
    c. V tach
    d. V Fib

2. What is recommended treatment of choice for patient who is hemodynamically stable?
    a. Clonidine
    b. Lisinopril with thiazides
    c. Beta blocker first for rate control, then amiodarone for rhythm control
    d. None of the above

An 18-year-old female is taken to ED from school because of a pulse that is 44 resting. She is a start athlete and marathon runner for the high school. She has no symptoms of shortness of breath or chest pain.

3. What test should be done?
    a. 12 lead EKG
    b. Stat Echo
    c. Cardiocentesis
    d. Cardiac massage

4. What is most appropriate treatment for this patient?
    a. No treatment needed
    b. OR stat
    c. IV heparin
    d. TPA

5. What medication may be given to a symptomatic patient with bradycardia to increase their pulse?
    a. metoprolol
    b. dobutamine
    c. Atropine
    d. cardene

6. What is the mechanism of action of atropine?
    a. Competitive inhibitor of muscarinic Ach receptors and increases firing through at SA node
    b. B2 agonist
    c. B1 agonist
    d. none of the above

34-year-old female presents to her PCP for routine check up. She has no complaints. EKG is shown below.

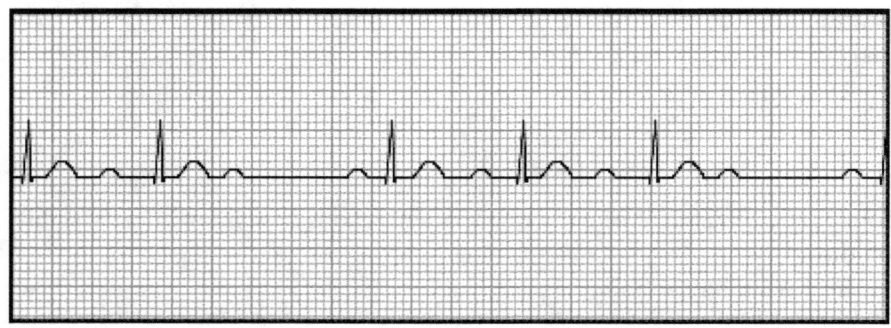

7. What is most likely diagnosis?
    a. 1st degree AV block
    b. 3rd degree AV block
    c. 2nd degree heart block Type 1
    d. none of the above

8. What is treatment of choice?
    a. No treatment needed
    b. OR immediately
    c. Chest tube
    d. Chest compressions

9. A type 2 mobitz heart block is treated with:
    a. Heart transplant
    b. atropine
    c. steroids
    d. Pacemaker

32-year-old female presents with shortness of breath. EKG is shown below.

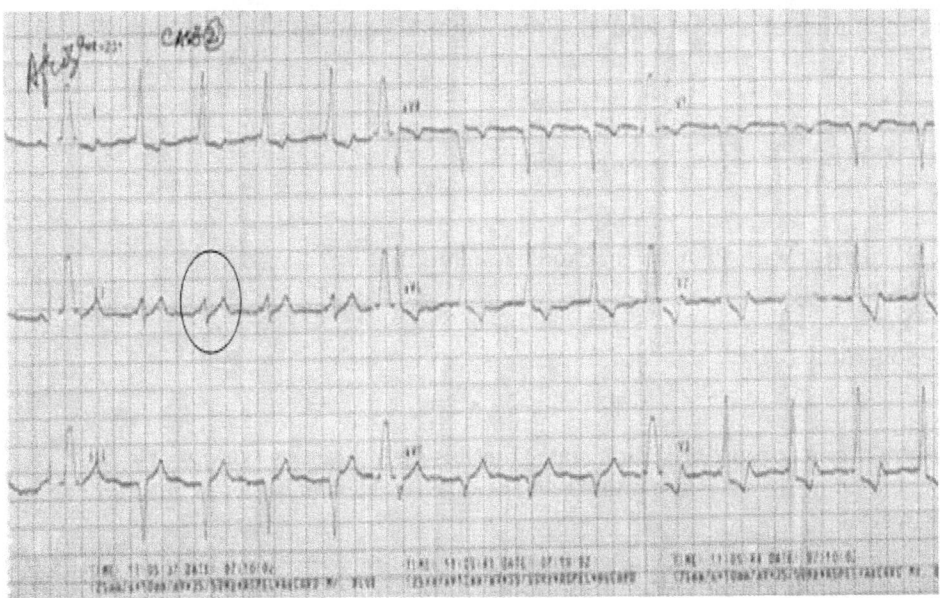

10. What is the most likely diagnosis?
    a. WPW syndrome
    b. Atrial flutter
    c. Atrial fibrillation
    d. V Tach

11. What is treatment of choice?
    a. Chest tube
    b. IV heparin
    c. Radioablation of Bundle of Kent
    d. Medrol dose pack

A 45-year-old male with end stage kidney disease is found to have electrical pulse activity but no palpable pulse.

12. What is most likely diagnosis?
    a. V Tach
    b. V fib
    c. PEA
    d. Atrial flutter

13. What are the typical causes of PEA?
    a. Hypovolemia/hypokalemia, hyperkalemia, hypoglycemia, acidosis, tension pneumothorax, trauma
    b. Hyperkalemia only
    c. Hyperkalemia and acidosis only
    d. Hyperkalemia, acidosis, and trauma only

14. After resuscitation of patient, what treatment should be administered urgently?
    a. CPR
    b. shock
    c. thoracocentesis
    d. Chest tube

15. Should the patient with PEA be shocked back into rhythm?
    a. NO
    b. Yes
    c. sometimes

A patient with chronic atrial fibrillation states that his vision is very yellow. He has occasional chest pains as well with nausea and vomiting. He states that he has not had his dialysis this week.

16. What is most likely diagnosis?
    a. He has been using illicit drugs
    b. He is Bipolar
    c. Digoxin toxicity
    d. He has a UTI

17. What is treatment of choice?
    a. Dialysis immediately, stop digoxin, digoxin antibodies
    b. OR immediately
    c. Stat cardiology consult
    d. Intubate immediately

18. How does digoxin work?
    a. B1 agonist
    b. B1 blocker
    c. inhibits release of dopamine from terminals
    d. Inhibits sodium-potassium ATPase channel

A 78-year-old female on amiodarone for 44 years for arrhythmias presents with shortness of breath. CXR shows opacities throughout both her lung fields.

19. What is most likely diagnosis?
    a. Pulmonary fibrosis
    b. Renal cancer
    c. Horners syndrome
    d. pneumonia

20. Amiodarone affect what phase of the cardiac cycle?

    a. 3
    b. 2
    c. 4
    d. 1

A 45-year-old male is treated for hypertension with clonidine.

21. What is the mechanism of action for clonidine?
    a. Central A2 Agonist
    b. B1 agonist
    c. B2 agonist
    d. Peripheral A1 blocker

22. This patient presents to ED with malignant hypertension in the 230 systolic. What is most likely cause for this?
    a. Not controlled on clonidine
    b. Was using drugs
    c. He stopped taking his clonidine and has rebound hypertension
    d. None of the above

A 56-year-old female treated for hypercholesterolemia on simvastatin presents with total body pain.

23. What labs should be obtained?
    a. Antiphospholipid antibody titer
    b. Type and screen
    c. Troponins
    d. CK, CBC, chem. 10, LFT

24. What is likely diagnosis with CK being very elevated?
    a. exercise
    b. Stress
    c. Statin-induced myalgias
    d. Aspirin-induced myalgias

25.

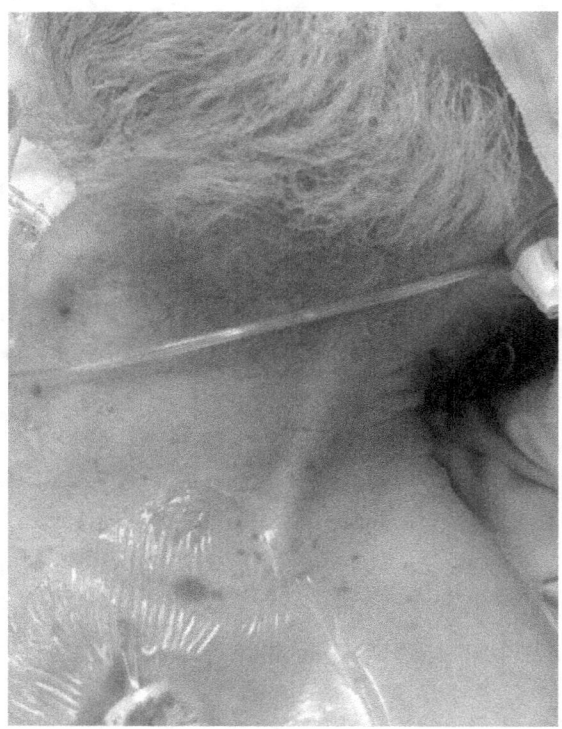

What pathology is noted here in this patient with ST segment elevation acutely requiring intubation and catheterization?

    a. Left JVD
    b. Left tracheal deviation
    c. Left central venous catheter hematoma
    d. pneumothorax

Answers: Cardiovascular

    1) Atrial Flutter

Atrial Flutter
-Irritable focus in the atria fire at 250 to 350 beats per minute
-Diagnose with EKG which shows sawtooth pattern of P waves
    -Treatment is similar to atrial fibrillation

    2) Beta blocker first for rate control, then amiodarone for rhythm control—see above
    3) 12 lead EKG
    4) No treatment needed
    5) Atropine

Sinus Bradycardia
-Heart rate less than 60 beats per minute
-Causes include:
    -Cardiovascular conditioning (i.e., athletes)
        -Normal
    -β blocker excess
-Asymptomatic bradycardia usually is not pathologic
-Symptomatic bradycardia warrants attention
-Diagnose with EKG which shows P wave before each QRS wave (i.e., sinus rhythm) with heart rate less than 100 (i.e., bradycardia)
-Treat underlying cause
-Consider atropine to increase heart rate
-Consider pacemaker placement in severe symptomatic bradycardia

    6) Competitive inhibitor of muscarinic Ach receptors and increases firing through at SA node
    7) 2nd degree heart block Type 1

-Type I Mobitz (i.e., Wenckebach)
    -PR interval continues to prolong precipitously over multiple beats until one beat is dropped (i.e., no QRS seen)
    -No treatment required

    8) No treatment needed
    9) Pacemaker
    10) WPW Syndrome

Wolff-Parkinson White Syndrome (WPW Syndrome)
-Accessory pathway from atria to ventricles causing premature ventricular excitation because of no delay in the atrioventricular (AV) node
    -Accessory pathway called the Bundle of Kent
-Impulse travels through AV node and depolarizes the ventricles, and then travels back through the accessory pathway and depolarizes the atria (i.e., reentry loop)

-Diagnose with EKG which shows a shortened PR interval, narrow complex (i.e., QRS interval less than 0.12 seconds or 120 milliseconds) tachycardia, and delta wave (i.e., an upward spike before the QRS complex)
-Treat with radioablation of the Bundle of Kent

11) Radioablation of Bundle of Kent
12) PEA

Pulseless Electrical Activity (PEA)
-Electrical activity detected on heart monitor but no pulse palpated
-Causes include:
- 6Hs
  - Hypoxia
  - Hypovolemia
  - Hypokalemia or hyperkalemia
  - Hypoglycemia
  - Hypothermia
  - Hydrogen ion (i.e., acidosis)
- 5Ts
  - Toxins
  - Tamponade
  - Trauma
  - Thrombosis
  - Tension pneumothorax
-Look for reversible causes (i.e., 6Hs and 5Ts) and treat underlying cause
-Initiate treatment with CPR
  -Not a shockable rhythm

13) Hypovolemia/hypokalemia, hyperkalemia, hypoglycemia, acidosis, tension pneumothorax, trauma
14) CPR
15) No
16) Digoxin toxicity

Digoxin
-Inhibits sodium-potassium ATPase channel
-Can be given to patients with atrial fibrillation for rhythm control
-Can also be used in the management of congestive heart failure (CHF)
-Narrow therapeutic window
-Side effects include:
  -Nausea
  -Vomiting
  -Vision turns to yellow
  -Hyperkalemia
  -Arrhythmias
-Hypokalemia exacerbates toxicity
-Check serum levels especially in those with renal failure
-If overdose suspected, can give digoxin antibodies (i.e., digibond, digoxin Fab)

17) Dialysis immediately, stop digoxin, digoxin antibodies

18) Inhibits sodium-potassium ATPase channel
19) Pulmonary fibrosis
20) 3
21) Central A2 Agonist
22) He stopped taking his clonidine and has rebound hypertension
23) CK, CBC, chem. 10, LFT
24) Statin induced myalgias

Statins
-Inhibit HMG-CoA reductase
    -Rate limiting step of cholesterol biosynthesis
-Reduces the production of LDL, which causes an increase in LDL receptor secondary to decreased negative feedback
    -Increased LDL receptor causes more LDL in the blood to be uptaken from the blood reducing serum LDL levels
-Useful in reducing LDL levels
-Side effects include:
    -Myalgias
        -Elevated creatine kinase (CK)
    -Elevated liver enzymes
    -Gastrointestinal (GI) side effects)
    -Potentiation of anticoagulant effect of warfarin

25. Left JVD

Pulmonary

45-year-old female presents with shortness or breath and skin lesions. She gets this every Spring. Her blood work is notable for elevated IgE. Her CXR is within normal limits. She has evidence of wheezing on physical exam.

1) What is the most likely diagnosis?
    a. Asthma
    b. Wegeners granulomatosis
    c. Goodpasture's Syndrome
    d. None of the above

2) What is treatment of choice?
    a. heparin

b. IVF
c. intubation
d. Bronchodilators, B2 agonists, anticholinergics

3) Zileuton is used to treat asthma as well. How does this medication work?
    a. Anti-inflammatory only
    b. Inhibits enzyme in lipoxygenase pathway, leukotriene inhibitors
    c. Steroid inhibitor
    d. Prevents mast cells from releasing histamine

4) Montelukast also can be used to treat asthma. How does this medication work?
    a. Prevents mast cells from releasing histamine
    b. Steroid inhibitor
    c. Internalizes the arachidonic acid receptors
    d. Leukotriene receptor blocker

5) Cromolyn is used occasionally in the treatment of asthma. What is the mechanism of action?
    a. Leukotriene receptor blocker
    b. Mast cell stabilizer
    c. Steroid inhibitor
    d. None of the above

6) What if the above patient does not get treatment for her acute asthma attack and starts to decompensate. Oxygen therapy does not help her. What is the diagnosis?
    a. Status asthmaticus
    b. Status epilepticus
    c. Panic attack
    d. Medication induced respiratory difficulty

7) What is immediate treatment of choice?
    a. Heparin gtt
    b. IVF
    c. IV steroids and epinephrine
    d. cromolyn

4-year-old male presents with shortness of breath. Liver enzymes are extremely high. Liver is palpated and is large. Patient has scleral icterus. Patients mother has this same problem.

8) What is likely diagnosis in this patient?
    a. A1 antitrypsin deficiency
    b. Potts Disease
    c. Goodpasture's Disease
    d. Wegeners Granulomatosis

9) What is the genetic inheritance?
   a. X Linked dominant
   b. X linked recessive
   c. Autosomal recessive
   d. Autosomal dominant

10) What structure is destroyed in this disease state?
    a. Elastase
    b. cartilage
    c. phospholipid bilayer
    d. none of the above

11) What is treatment of choice?
    a. O2 only
    b. A1 antitrypsin from serum only
    c. O2 and bronchodilators only
    d. Giving A1 antitrypsin to patient from pooled human serum, O2, bronchodilators

12) If these treatments do not work, what is long term treatment option?
    a. Bronchodilators
    b. steroids
    c. Liver and Lung transplant
    d. No definitive treatment

34-year-old male with infertility and diabetes presents with shortness of breath. CXR shows a left upper lobe consolidation. PFT show obstructive picture. This patient has had recurrent pneumonias in the past.

13) What is likely diagnosis?
    a. Goodpastures
    b. Cystic fibrosis
    c. Wegeners Granulomatosis
    d. Chronic Bronchitis

14) What is inheritance pattern?
    a. Autosomal recessive
    b. Autosomal dominant
    c. X linked recessive
    d. None of the above

15) What chromosome is involved?
    a. 7

b. 8
c. 9
d. 4

16) What gene is disabled in CF patients?
    a. Potassium channel
    b. ATPase gene
    c. AMP gene
    d. Chloride channel-CFTR

17) What can be done to screen for CF?
    a. Blood work
    b. CT head and CXR
    c. Prenatal screening with amnio
    d. Nothing can be done

18) What is the most common mutation in CF?
    a. Delta 508
    b. Delta 408
    c. Delta 500
    d. Delta 504

19) What organism is most common cause of pneumonia in CF patients?
    a. Klebsiella
    b. Pseudomonas Aeuruginosa
    c. mycoplasma
    d. streptococcus

20) What is treatment of choice for the pneumonia?
    a. Zosyn
    b. penicillin
    c. vancomycin
    d. Ciprofloxacin

56-year-old female s/p right lung lobectomy due to small cell lung cancer presents with shortness of breath. She is receiving radiation therapy. CXR shows ground glass appearance on left side and diffuse infiltrates. No fever is documented and wbc count is normal.

21) What is likely diagnosis?
    a. MI
    b. pneumonia
    c. Radiation pneumonitis
    d. None of the above

22) What is treatment of choice?
   a. Steroids
   b. NSAIDS
   c. Albuterol
   d. Atrovent

34-year-old male with 15 year history of smoking presents with a black left 4th finger tip. Denies trauma.

23) What syndrome is most likely?
   a. Kawasaki Disease
   b. Buergers Disease
   c. Behcet Syndrome
   d. Takahashi syndrome

24) What counseling should be administered to patient?
   a. Stop smoking
   b. No Fattening foods
   c. Decrease exercise to 4 hours per day
   d. None of the above

25) What is treatment for this patient?
   a. Heparin drip
   b. Amputate the gangrenous digit
   c. Bicarbonate drip
   d. Coumadin for life

56-year-old female presents with shortness of breath. CXR shows calcified linear plaques along the diaphragm. She has a history of shipbuilding and construction many years ago. CT chest shows pleural plaques.

26) What is likely diagnosis?
   a. Buergers disease
   b. Kawasaki Disease
   c. Pleural effusions
   d. Malignant mesothelioma

27) This disease is associated with exposure to what two compounds?
   a. Asbestos and smoking
   b. Lead
   c. Mercury
   d. Silver

28) What is next option in management of this patient?
   a. OR immediately to remove the lung

b. Biopsy lung
c. Start radiation
d. None of the above

29) Biopsy will show what finding:
   a. Lead lines
   b. Psammoma bodies
   c. Asbestos particles or ferruginous bodies.
   d. Giant cell Macrophages with schistocytes

Answer: Pulmonary

1) Asthma

Asthma
- Airway hypersensitivity
- Smooth muscle hyperactivity to constrict
- Wheezing
- Shortness of breath
- Associated with atopic dermatitis and eczema
- Eosinophilia
- Elevated IgE
- Symptoms and treatment dependent on peak flow and nighttime symptoms
- Respiratory alkalosis secondary to tachypnea, but when their serum pH normalizes, beware that they might be tiring out and entering respiratory failure
    - Beware of need to intubate
- Treatment includes bronchodilators including β2 agonists (i.e., albuterol), steroids, or anticholinergics (i.e., ipratropium)
    - Provide leukotriene inhibitors as modifying agents (i.e., zileuton, montelukast) to lower dose of steroid required
        - Zilueton
            - Inhibits enzyme in lipoxygenase pathway
        - Montelukast
            - Leukotriene receptor blocker
    - Less often utilized asthma medications include:
        - Cromolyn sodium
            - Mast cell stabilizer
            - Used as prophylaxis for exercise
        - Theophylline
            - Increases cAMP
            - Relaxes smooth muscle
            - Has a narrow therapeutic window
            - Contraindicated in pregnancy
- Avoid asthma triggers
- Give influenza shot yearly
- Give vaccines
- Stop smoking

2) Bronchodilators, B2 agonists, anticholinergics—see above
3) Inhibits enzyme in lipoxygenase pathway, leukotriene inhibitors
4) Leukotriene Receptor Blocker
5) Mast cell stabilizer
6) Status Asthmaticus

Status Asthmaticus
- Acute asthma exacerbation where standard bronchodilator therapy and oxygen not helpful
- Patient who does not get treated for acute asthma exacerbation who can decompensate further into respiratory failure
- Signs and symptoms include:
    - Wheezing

                -Cyanotic
                -Severe shortness of breath
-Treat with ABCs
-Start IV steroids
-Give epinephrine

    7) IV steroids and epinephrine
    8) A1 antirypsin deficiency

α-1 Antitrypsin (A1AT) Deficiency
-Autosomal recessive
-Deficiency in α-1 antitrypsin which usually inhibits elastase
-No inhibition of elastase leads to destruction of elastic tissue
        -Lung becomes less elastic and more compliant
-Classic history is child presenting with cirrhosis and all its signs and symptoms
-Causes panlobular emphysema
        -Versus centrilobular emphysema caused by smoking
-Consider in patients with rapid deterioration in lung functioning without smoking history
-Elevated liver enzymes
-Low α-1 antitrypsin level in the blood
-PFTs show obstructive lung disease picture
-Definitive diagnosis is through genetic testing
-Treat by giving α-1 antitrypsin to patients from pooled human serum
-Give oxygen and bronchodilators for exacerbations
-Emphasize smoking cessation
-Give influenza shot yearly
-Give vaccines
-Consider lung and liver transplant

    9) Autosomal Recessive—see above
    10) Elastase
    11) Giving a1 antirypsin to patient from pooled human serum, O2, bronchodilators
    12) Liver and Lung transplant
    13) Cystic Fibrosis

Cystic Fibrosis (CF)
-Autosomal recessive
-Mutation is on chromosome 7
-Mutation is in chloride channel (i.e., CFTR)
-More common in Caucasians
-Prenatal screening with amniocentesis possible
-Many different mutations in same gene results in the disease
        -Most common mutation is ΔF508 (i.e., deletion of 3 base pairs)
        -Newborn screening detects ΔF508 mutation
-Complications include:
        -Infertility
                -Problem with sperm motility
        -Pancreatic damage
                -Diabetes mellitus
                -Malabsorption of fat-soluble vitamins (i.e., Vitamins A, D, E, and K)

- Pancreatic damage secondary to lack of secretion of fluid producing proteinaceous plugs resulting in the backup of pancreatic enzymes that eventually degrade the pancreas itself
  - Meconium ileus
    - In newborns, diagnose and treat with gastrograffin enema
  - Recurrent lung infections
    - Pseudomonas
      - Most common cause of pneumonia in CF patient
    - Staphylococcus aureus
    - B. sepacia
    - Lung damage secondary to lack of ability to clear airways of bacteria given lack of fluid and defective mucociliary escalator
      - Mucus plugs and obstruction present contributing to recurrent infections
  - Bronchiectasis
    - Secondary to recurrent infections
  - Liver cirrhosis
- Classic history is newborn that may taste salty
- PFTs show obstructive lung disease picture
- Gold standard in diagnosis is sweat chloride test (i.e., iontophoresis)
  - Sodium chloride (NaCl) above 70 is diagnostic
- Treat with antibiotics that include piperacillin-tazobactam (i.e., Zosyn) or gentamicin to cover Pseudomonas and gram negative rods
- Give influenza vaccine yearly
- Give vaccines
- Administer chest physiotherapy (PT)
- Pancreatic enzyme supplements to absorb fat-soluble vitamins

14) Autosomal recessive
15) 7
16) Chloride channel-CFTR
17) Prenatal screening with Amnio
18) Delta 508
19) Pseudomonas Aeuruginosa
20) Zosyn
21) Radiation pneumonitis

Radiation Pneumonitis
- Interstitial inflammation of the lung post-radiation of the chest
- CT shows ground-glass appearance and diffuse infiltrates
- PFTs show restrictive lung disease picture
- Treatment is corticosteroids

22) Steroids

23) Buergers Disease

Thrombangiitis Obliterans (i.e., Buerger's Disease)
- Seen in males who smoke
- Thrombosis in the digits leading to gangrene
- Treat by stop smoking

24) Stop smoking
25) Amputate the gangrenous digit
26) Malignant mesothelioma

Malignant Mesothelioma
-Associated with exposure to asbestos
-Associated with smoking
-Related to occupational history of shipbuilding and working in construction with older buildings
-Workup should begin with a chest x-ray which may show calcified linear plaques along the diaphragm
-CT should follow to help in diagnosis which may show pleural plaques
-Histology reveals asbestos particles that look like a guitar (i.e., ferruginous bodies)
-Poor prognosis as survival is a few months after diagnosis

27) Asbestos and smoking
28) Biopsy Lung
29) Asbestos particles or ferruginous bodies.

Hematology-Oncology

35-year-old male presents to ED with fatigue, shortness of breath, easy bruising and bleeding from nose. He has been battling pneumonias every 5 months.

1) What lab will be helpful in the diagnosis?
    a. CBC
    b. PTT
    c. PT
    d. Magnesium

2) What will the peripheral smear show?
    a. Bite cells
    b. schistocytes
    c. Auer rods
    d. Target cells

3) What pathology does this patient have?
    a. AML
    b. ALL
    c. CLL
    d. ATL

4) Definitive diagnosis can be obtained by:
    a. Lung biopsy
    b. Brain Biopsy
    c. Bone marrow biopsy
    d. Peripheral smear

5) What consultation should be placed?
    a. Hematology-Oncology
    b. Neurosurgery
    c. Pediatrics
    d. Family Medicine

6) What lab test differentiates AML from Acute Lymphocytic Leukemia (ALL)?
    a. Positive myeloperoxidase
    b. ATL receptor status
    c. Bite cells
    d. Target cells

8-year-old male with fatigue, fever, UTI, splenomegaly, and large lymph nodes is seen. CBC is done and shows multiple blasts.

7) What is most likely diagnosis?

a. ALL
   b. AML
   c. CLL
   d. CML

8) What leukemia has the highest cure rate at 85%?
   a. AML
   b. CLL
   c. CML
   d. ALL

9) What is treatment of choice?
   a. Chemotherapy
   b. XRT
   c. Surgery
   d. No Treatment exists

10) ALL is increased in what diseases?
    a. Downs, fanconi's anemia, brutons agammaglobulinemia
    b. Downs only
    c. Fanconi's anemia only
    d. Downs and brutons only

55-year-old male presents with fever and easy bruising and bleeding. On exam he has splenomegaly. CBC shows increase WBC.

11) What is likely diagnosis?
    a. CML
    b. CLL
    c. ALL
    d. AML

12) What procedure is key to the diagnosis of CML?
    a. Karyotyping—Chrom 9 and 22 translocation
    b. Peripheral smear
    c. CT head
    d. CT bone marrow

13) What receptors are encoded on these translocations?
    a. ATR gene upregulation
    b. CFTR gene upregulation
    c. BCR-ABL gene=tyrosine kinase upregulation
    d. None of the above

14) What is the terminology when the myeloblasts increase greater than 20 % in these patients with CML?
   a. Blast crisis
   b. DIC
   c. Pulmonary embolus
   d. Tumor phenomenon syndrome

15) What is standard treatment for CML?
   a. Imatinib(Gleevac)
   b. XRT
   c. Aspirin
   d. Lovenox

16) How does Imatinib work?
   a. CFTR gene dysregulator
   b. Tyrosine kinase inhibitor
   c. AMPA channel upregulation
   d. None of the above

75-year-old man with splenomegaly on exam and pancytopenia on CBC. Peripheral smear shows many mononuclear cells with abundant cytoplasmic projections.

17) What is diagnosis?
   a. Non small cell lung cancer
   b. Small cell lung cancer
   c. AML
   d. Hairy cell leukemia

18) What stain is used to diagnose this disease?
   a. TRAP
   b. FLAP
   c. PLAP
   d. BHCG

19) What is treatment in symptomatic patient?
   a. IVF only
   b. Lovenox
   c. Coumadin and lovenox
   d. Nucleoside analogs

55-year-old male with history of lymphoma currently being treated with chemotherapy presents with shortness of breath and difficulty voiding. Labs show hyperkalemia, hyperphosphatemia, and Hyperuricemia. Calcium level is low.

20) What is diagnosis?
   a. DIC

b. Tumor lysis syndrome
  c. End stage liver Disease
  d. None of the above

21) What is treatment?
  a. Aspirin
  b. Gentamicin
  c. Allopurinol
  d. Dulcolax

45-year-old male presents with night sweats and dry cough and itchy skin. He has cervical adenopathy. He has CXR showing left hilar adenopathy. His LDH level is elevated.

22) What must be obtained next for diagnosis?
  a. Excisional lymph node biopsy
  b. Lung biopsy
  c. Sputum sample
  d. None of the above

23) What cells are often seen in patient with HL?
  a. Reed-Sternberg cells
  b. Owl eye cells
  c. Schistocytes
  d. target cells

24) What is treatment?
  a. OR stat
  b. Chemo and XRT
  c. XRT only
  d. None of the above

25) What is common side effect from cisplatin?
  a. Nephrotoxic and ototoxic
  b. Gallstones
  c. Pulmonary fibrosis
  d. Lung cancer

26) What is most common side effect from bleomycin?
  a. Dry skin
  b. Interstitial nephritis
  c. Pulmonary fibrosis
  d. Stevens Johnson Syndrome

34-year-old male in hospital for pneumonia. CBC shows Hematocrit of 22. While patient is being transfused, he has a fever to 40 C.

27) What is next step in management?
    a. Start IVF
    b. Give benadryl only
    c. Stop transfusion, give Tylenol
    d. Nothing can be done at this point

28) What if patient starts to itch with fever?
    a. Stop transfusion, Tylenol, antihistamines
    b. Give Benadryl
    c. Give Epi
    d. Do not change anything

29) What immunoglobulin is associated with the reaction just above?
    a. IgA
    b. IgG
    c. IgM
    d. IgD

30) What if patient becomes hypotensive and tachycardic?
    a. Just give benadryl
    b. Just give Epi
    c. Stop transfusion, IVF, vasopressors if needed
    d. Do not change anything, the reaction will resolve spontaneously

45-year-old male with loss of proprioception while walking occurring over the last 3 months. He had a prior abdominal surgery for small bowel tumor.

31) What lab work should be done?
    a. CBC-macrocytic anemia, vitamin b12 level, methionine level, homocysteine level
    b. Vit B12 and methionine level only
    c. CBC only needed
    d. Homocysteine and methionine only

32) What is the most likely diagnosis?
    a. Niacin deficiency
    b. Folate deficiency
    c. Vitamin b12 deficiency
    d. None of the above

33) What is cause of this patient's spinal cord pathology?
    a. Frederick's ataxia
    b. Subacute combined degeneration
    c. Syringomyelia
    d. Marcus Gunn syndrome

34) What blood levels are elevated in this disease?
    a. Vitamin B12
    b. Riboflavin
    c. Methionine, homocysteine
    d. Methionine only

35) What is treatment of choice?
    a. Vit b12 and Folate together
    b. Folate only
    c. Vit b12 only
    d. niacin

Answers: Hematology-Oncology

1) CBC-shows leukocytosis, thrombocytopenia, and anemia

Acute Myelogenous Leukemia (AML)
-Most common acute leukemia affecting adults
-Incidence increases with age
-Signs and symptoms include:
- -Fatigue
- -Shortness of breath
- -Easy bruising and bleeding
- -Infection
- -Splenomegaly
-Risk factors include:
- -Ionizing radiation
- -Chemotherapy
- -Genetics
-Diagnose initially with complete blood count (CBC)
- -Leukocytosis
- -Thrombocytopenia
- -Anemia
-May see Auer rods in peripheral smear
- -Red rod shaped structures in the cytoplasm of myeloblasts
-Definitive diagnosis requires bone marrow biopsy
- -Greater than 20 % of the bone marrow infiltrated with myeloblasts
-Positive myeloperoxidase
- -Differentiates AML from Acute Lymphocytic Leukemia (ALL)
-Treatment involves chemotherapy with cytarabine
-Bone marrow transplant is best treatment overall for remission
-Several subtypes
- -Acute Promyelocytic Leukemia (APL)
- -Translocation of chromosomes 15 and 17
- -Treat with trans-retinoic acid

2) Auer rods-see above
3) AML-see above
4) Bone marrow biopsy
5) Hematology-Oncology
6) Positive Myeloperoxidase
7) ALL

Acute Lymphocytic Leukemia (ALL)
-Most common cancer in childhood (i.e., 4-12 years)
-Good prognosis (i.e., cure rate 85 %)
-Increased incidence in:
- -Down syndrome
- -Fanconi's anemia
- -Ataxia-telangiectasia
- -Bruton's agammaglobulinemia
-Signs and symptoms include:

                -Fatigue
                -Fever
                -Infections
                -Enlarged lymph nodes
                -Splenomegaly
                -Petechiae
-Diagnose initially with CBC
        -Note blasts
-Peripheral smear shows blasts
-Definitive diagnosis made with bone marrow biopsy definitive
        -Shows blasts
-Treat with chemotherapy

8) ALL
9) Chemotherapy
10) Downs, fanconi's anemia, brutons agammaglobulinemia
11) CML

Chronic Myelogenous Leukemia (CML)
-Signs and symptoms include:
        -Fatigue
        -Fever
        -Gout
        -Infections
        -Easy bruising and bleeding
        -Splenomegaly
-Often asymptomatic at diagnosis
-Can present with incidental increase in white blood cell count on routine CBC
-Philadelphia chromosome is key in diagnosis
        -Chromosomes 9 and 22 translocation
                -Bcr-abl fusion
                        -Unregulated tyrosine kinase
-Bone marrow biopsy often not necessary
-CML must be distinguished from leukemoid reaction
        -Leukemoid reaction has positive leukocyte alkaline phosphatase
        -CML negative for leukocyte alkaline phosphatase
-Can evolve into a blast crisis
        -Greater than 20 % myeloblasts in the blood or bone marrow
-Treat with imatinib (i.e., Gleevac), a tyrosine kinase inhibitor

12) Karyotyping—Chrom 9 and 22 translocation
13) BCR-ABL gene=tyrosine kinase upregulation
14) Blast Crisis
15) Imatinib(Gleevac)
16) Tyrosine kinase inhibitor
17) Hairy Cell Leukemia

Hairy Cell Leukemia

-Disorder of B cells
-More commonly in older men
-Splenomegaly
-Pancytopenia
-Bone marrow infiltration
-Peripheral smear shows hairy cells
- -Mononuclear cells with abundant pale cytoplasm and cytoplasmic projections
-Diagnosis confirmed by identifying hairy cells in the blood, marrow, or spleen
-Tartrate-resistant acid phosphatase (TRAP) staining of hairy cells
-Treat if symptomatic with nucleoside analogs

    18) TRAP
    19) Nucleoside analogs
    20) Tumor Lysis Syndrome

Tumor Lysis Syndrome
-Seen in gout and in patients with leukemia or lymphoma treated with chemotherapy;
-Characteristics:
- -Hyperkalemia
- - -Can cause cardiac conduction abnormalities and muscle weakness
- -Hyperphosphatemia
- - -Can cause respiratory depression
- -Hyperuricemia
- - -Can cause uric acid nephropathy
- -Hypocalcemia
-Treat with allopurinol (i.e., to decrease production of uric acid by inhibiting xanthine oxidase)
- -May give to patients who are being treated with chemotherapy as prophylaxis against tumor lysis syndrome

    21) Allopurinol
    22) Excisional Lymph Node Biopsy
    23) Reed-Sternberg cells-Giant abnormal B cells with bilobar nuclei and high eosinophilic nucleoli that create an owl's eye appearance
    24) Chemo and XRT
    25) Nephrotoxic and ototoxic
    26) Pulmonary Fibrosis

Chemotherapeutic Drug Reactions
-Cyclophosphamide – hemorrhagic cystitis
-Cyclosporine (i.e., blocks expression of IL-2 gene) – nephrotoxicity
-Cisplatin – nephrotoxicity, ototoxicity
-Doxorubicin – dilated cardiomyopathy
-Vincristine – neuropathy
-Bleomycin – pulmonary fibrosis
-Methotrexate – folic acid deficiency, liver disease, nephrotoxicity, intracranial inflammation

    27) Stop transfusion, give Tylenol
    28) Stop transfusion, Tylenol, antihistamines

Transfusion Reactions
-1 unit of RBC transfusion increases Hg by 1 and thus Hct by 3 percentage points
-Nonhemolytic febrile transfusion reaction
- Secondary to cytokines during storage of blood
- Patient becomes immediately febrile
- Stop transfusion
- Give Tylenol
-Minor allergic reaction following transfusion
- Involves antibody formation (i.e., usually IgA) against donor proteins
- Usually occur following transfusion of plasma containing product
- See urticaria
- Give antihistamines
- If severe reaction (i.e., anaphylactic), stop transfusion and give epinephrine
-Hemolytic transfusion reaction
- Secondary to development of antibodies against donor erythrocytes (i.e., ABO incompatibility)
- Signs and symptoms include:
  - Fever
  - Flushing
  - Burning at IV site
  - Tachycardia
  - Tachypnea
  - Hypotension
- Signs and symptoms begin after merely a small amount of blood transfused
- Stop transfusion
- Replace donor blood with normal saline
- Administer vigorous IV fluids
  - Consider vasopressors if needed
  - Monitor urine output

29) IgA
30) Stop transfusion, IVF, vasopressors if needed—see above
31) CBC-macrocytic anemia, vitamin b12 level, methionine level, homocysteine level
32) Vitamin b12 deficiency
33) Subacute combined degeneration
34) Methionine, homocysteine
35) Vit b12 and Folate together

Vitamin B12 versus Folic Acid Deficiencies
-Vitamin B12 deficiency
- B12 stored in liver for 5 years
  - Takes a long time to become deficient
- Get B12 from animal proteins
- Get neurological symptoms
  - Subacute combined degeneration
    - Dorsal column or posterior column degeneration with lateral column involvement
    - Positive Romberg sign

- Decreased vibration and proprioception
- Low serum B12
- Elevated methionine
    - More sensitive than serum B12 level
- Elevated homocysteine levels
    - More sensitive than serum B12 level
- Folate deficiency
    - Folic acid present in green vegetables
    - Deficiency alone will not cause neurological symptoms
    - Low serum folate
    - Normal methionine
    - Elevated homocysteine
- Note multilobed, hypersegmented neutrophils on peripheral smear in either deficiency
    - Hypersegmented as these cells are immature and are released given lack of folic acid and B12 required for cell division and DNA replication
- In patients with questionable B12 or folate deficiency, do not give B12 alone as you will make neurologic symptoms worse
    - Give both B12 and folate

Gastrointestinal

24-year-old male studying for the boards notices shortness of breath. He feels a sharp burning feeling over his chest and tastes sour substances when he is sleeping at night.

1) What is most likely diagnosis?
   a. GERD
   b. Paraesophageal hernia
   c. Zolliger Ellison Syndrome
   d. MEN 1

2) Which of the following is a risk factor for the above pathology?
   a. Gum
   b. H.Pylori
   c. Milk
   d. Vitamin B12

3) What substance should he avoid?
   a. Caffeine
   b. Sugar
   c. Milk
   d. Caramel

4) What is treatment of choice?
   a. NSAIDS
   b. Drinking lots water
   c. Multivitamins
   d. H2 blocker

5) What complication may occur if untreated?
   a. Barrett's Esophagus
   b. Colonic ulcer
   c. Meckels Diverticulum
   d. Esophageal hernia

45-year-old male with AIDS presents with chest pain. He has had fevers and chills and has been having swallowing difficulty solids.

6) What must be ruled out?
   a. Hernia
   b. Infectious Esophagitis
   c. Gangrenous cholecystitis
   d. Duodenal ulcer

7) Which of the following is the most likely cause in this patient?
   a. CMV
   b. HIV
   c. HHPV
   d. Candida

8) What is the best diagnostic modality?
    a. Upper Endoscopy
    b. colonoscopy
    c. CXR
    d. CT abdomen

9) What is treatment?
    a. IV gentamicin
    b. PO gentamicin
    c. Oral Fluconazole
    d. None of the Above

10) What if on endoscopy, a sample from the esophagus is biopsied and pathology shows intranuclear inclusion bodies with owl eyes. What is the diagnosis?
    a. CMV
    b. HPV
    c. HSV
    d. Molluscum contagiousum

11) What is treatment for this condition?
    a. NSAIDS
    b. Requip
    c. Ganciclovir
    d. Copaxone

45-year-old female presents with difficulty swallowing solids and liquids. She has had 25 pounds of weight loss.

12) What study should be done next?
    a. Barium swallow
    b. CT Pelvis
    c. Rectal Exam
    d. Plasma agglutination levels

13) Barium swallow reveals bird beak sign of the esophagus. What is the cause for this condition?
    a. Ulcer esophagus
    b. Infection of the esophagus
    c. Congenital defect in upper GI
    d. Loss of inhibitory interneurons

14) What is diagnosis?
    a. Achalasia
    b. Hirshprung's Disease
    c. Chagas Disease
    d. None of the above

15) What does manometry testing reveal?
   a. Increased LES contractility and no peristalsis
   b. Decreased LES contractility
   c. Minimal LES contractility
   d. No abnormalities

16) What is treatment in short term?
   a. NSAIDS
   b. Resect esophagus
   c. Radiate the esophagus
   d. CCB or botox

17) What is long-term treatment?
   a. Myotomy or dilation
   b. Botox
   c. Resection Esophagus
   d. Radiate the esophagus

18-year-old woman presents with difficulty swallowing. She is noted to be very pale with very low energy level.

18) What lab should be sent?
   a. CBC
   b. INR
   c. Glomerular filtration level
   d. Hepatitis B antibodies

19) CBC shows iron deficiency anemia. What is likely diagnosis?
   a. Johnson syndrome
   b. Plummer-Vinson Syndrome
   c. Tabes Dorsalis
   d. Wernicke Syndrome

20) What is the pathology in this syndrome?
   a. Esophageal webs
   b. Diffuse esophageal spasm
   c. Achalasia
   d. Boerhaave's Syndrome

Answers: Gastointestinal

1) GERD

Gastroesophageal Reflux Disease (GERD)
- Signs and symptoms include:
  - Burning chest pain
  - Cough
  - Shortness of breath
  - Sore throat
  - Water brashing
  - Sour taste in the mouth after waking up
- Chest pain worse:
  - With fatty foods
  - 2 hours after eating
  - While lying down
- Pain relieved by antacids
- Most commonly secondary to reduction in lower esophageal sphincter (LES) tone
- Risk factors include:
  - Helicobacter pylori
  - Sliding hiatal hernia
    - LES herniates through the diaphragm
    - Resulting in a lack of assistance from diaphragm in producing LES pressure to prevent reflux
    - May consider leaving alone
    - Treat with Nissen fundoplication if high risk of cancer from reflux
  - Increased intraabdominal pressure
    - Obesity
    - Pregnancy
    - Valsalva maneuver
  - Food that reduces LES pressure
    - Chocolate
    - Caffeine
    - Mints
  - Sleeping soon after eating (i.e., less than 30 minutes after eating)
  - Lying down soon after taking bisphosphonates
  - Scleroderma
  - Zollinger-Ellison (ZE) syndrome
- Clinical diagnosis
- Consider treatment with H2 blocker first
  - Reversibly blocks the H2 receptor on parietal cells that stimulate release of acid
  - Cheaper and less effective with more side effects than proton pump inhibitors (PPIs)
    - Side effects include:
      - Hyperprolactinemia
      - Galactorrhea
      - Gynecomastia
      - P450 inhibition
- Also consider lifestyle changes including weight loss and lying down more than 30 minutes after eating
- If not responsive, consider PPI

-Irreversibly blocks the H/K ATPase, reducing the release of hydrogen ion from the gastric parietal cell
-More expensive and more effective with less side effects than H2 receptor blockers
-Complications include:
- -Esophagitis
- -Barrett's esophagus
    - -Precursor to esophageal adenocarcinoma
- -Stricture
- -Upper gastrointestinal (UGI) bleeding

2) H.Pylori
3) Caffeine
4) H2 blocker
5) Barrett's Esophagus
6) Infectious Esophagitis

Infectious Esophagitis
-Seen more commonly in immunosuppressed patients (i.e., HIV/AIDS)
-Causes include:
- -Candida
    - -Oral thrush
    - -Diagnose with KOH preparation
    - -Treat with oral fluconazole
- -Herpes simplex virus (HSV)
    - -Oral vesicles
    - -Diagnose with Tzanck smear showing multinucleated giant cells
    - -Treat with oral acyclovir
- -Cytomegalovirus (CMV)
    - -Causes retinitis
    - -Diagnose with smear showing intranuclear inclusion bodies with owl's eye appearance
    - -Treat with oral ganciclovir

7) Candida
8) Upper Endoscopy
9) Oral Fluconazole
10) CMV
11) Ganciclovir
12) Barium swallow
13) Loss of inhibitory interneurons
14) Achalasia

Achalasia
-Motility disorder
-Caused by loss of inhibitory interneurons in Auerbach's plexus
-Dysphagia to solids and liquids
-Weight loss may be present
-Secondary causes include:
- -Chagas disease

-Malignancy (i.e., pseudoachalasia)
    -Gastric cancer
    -Lymphoma
-Barium swallow reveals bird beak sign
-Manometry reveals increased LES contractility (i.e., lack of relaxation) and lack of peristalsis
-Treatment in short term for relief is with nitrates, CCBs, or botulinum toxin to relax LES
-Pneumatic dilation or surgical myotomy for long term relief

15) Increased LES contractility and no peristalsis
16) CCB or Botox
17) Myotomy or dilation
18) CBC
19) Plummer-Vinson Syndrome

Plummer Vinson Syndrome
-More common in women
-Dysphagia
-Iron deficiency anemia with esophageal webs
-Increased risk of squamous cell carcinoma of the esophagus

20) Esophageal webs

Infectious Disease

24-year-old female presents with fevers and altered mental status. She has had a witness seizure as well. CT head is negative. All of her labs are within normal limits.

1. What is next step in management?
    a. LP
    b. OR immediately
    c. Place shunt
    d. Place lumbar drain

2. If an LP is done and it shows normal glucose, elevated wbc, elevated protein, and negative gram stain, what is the likely diagnosis?
    a. Pseudotumor Cerebri
    b. Viral encephalitis/meningitis
    c. CSF Leak
    d. None of the above

3. Which of the following is a common cause of viral encephalitis?
    a. HSV
    b. CMV
    c. HPPV
    d. HCV

4. What is treatment of choice?
    a. Copaxone
    b. Robaxin
    c. Acyclovir
    d. Sulfanomides

5. What is the most common cause of meningitis from 0-3 months?
    a. Group B Strep
    b. Group C Strep
    c. Group D Strep
    d. Group A Strep

24-year-old male with HIV presents with fever and seizures. He stated that he has been around birds lately as he has a strong passion to save as many birds as possible from being hurt.

6. What study should be done first?
    a. LP
    b. EEG
    c. EMG/NCS
    d. CT head

7. LP is done. KOH prep shows budding yeast. What stain is used to show a halo due to thick capsule?
   a. Trichrome stain
   b. H and E stain
   c. India Ink
   d. None of the above

8. What is the Diagnosis?
   a. Cryptococcus
   b. Coccidiomycosis
   c. Blastomycosis
   d. Histoplasmosis

9. What is treatment?
   a. NSAIDS
   b. Isolation with IVF
   c. Tetracycline
   d. Flucytosine and Amp B

A 67-year-old female states that she has had severe headaches and nausea and vomiting recently after he trip to the lake. She has taken a swim in the lake and stated the water was very warm.

10. What is the likely diagnosis?
    a. Naegleria fowleri
    b. H influenza
    c. Paragonomiasis
    d. Chagas Disease

11. Through what route do these organisms enter the brain?
    a. Mouth
    b. Ears
    c. Skin
    d. Olfactory mucosa

56-year-old female with uncontrolled diabetes presents with severe headaches and fevers. She has a large and erythematous right eye with a bruit heard over that eye.

12. What is likely cause of this disorder?
    a. Mucor
    b. HIV
    c. Nocardia
    d. Giardia lamblia

13. What syndrome does the patient have?
    a. ICA occlusion
    b. Vertebral artery dissection
    c. Cavernous sinus Thrombosis
    d. Stroke

14. What is seen on histopathology?
    a. Non septate hyphae
    b. Septate Hyphae
    c. Yeast only
    d. Yeast with thick capsule only

15. What is treatment?
    a. Antibiotics only
    b. IVF and Antifungals only
    c. Emergency surgical debridement and Amp B
    d. No treatment available

34-year-old male with HIV presents with fever and nonproductive cough. CXR shows diffuse hilar infiltrates. Sputum culture is obtained and silver stain reveals the organism.

16. What is the diagnosis?
    a. Pneumocystis carinii
    b. klebsiella
    c. giardia
    d. aspergillosis

17. ABG is done showing P02 of 55. What should be given next?
    a. Robaxin
    b. NSAIDS
    c. Steroids
    d. Dulcolax

18. What is treatment?
    a. Bactrim
    b. Gentamicin
    c. Flucytosine
    d. Amphotericin B

3-month-old child has watery diarrhea. She has just started daycare.

19. What is the diagnosis?
    a. Apergillosis
    b. Rotavirus
    c. Giardia
    d. None of the above

20. What type of virus is this?
    a. Double stranded RNA
    b. Single stranded DNA
    c. Single stranded RNA
    d. Double stranded DNA

Answers: Infectious Disease

1. LP
2. Viral encephalitis/meningitis

Viral Encephalitis
-Infection of the parenchyma
-Causes include:
- -HSV
- -CMV
- -EBV
- -Enteroviruses
- -Arboviruses (i.e., West Nile virus)
- -Rabies virus

-Signs and symptoms include:
- -Fever
- -Altered mental status
- -Seizures
- -Non-focal neurologic deficits

-CT to rule out mass lesion and hemorrhage
-LP shows:
- -Normal glucose
- -Elevated WBCs (i.e. lymphocytes)
- -Elevated protein
- -send CSF for Gram Stain, Acid-Fast, India Ink, PCR

-Treat HSV encephalitis with acyclovir
-Treat CMV encephalitis with ganciclovir
-Treatment for all other causes of viral encephalitis is usually not necessary as they are self-limited

3. HSV
4. Acyclovir
5. Group B Strep
6. CT head
7. India Ink
8. Cryptococcus
9. Flucytosine and Amp B
10. Naegleri fowleri

Naegleria fowleri
-Protozoa
-Found in rivers, lakes, and warm waters
-Can enter through the nostrils, invade through the olfactory mucosa or bulb, enter the cribiform plate, and then enter the CNS
-Causes a fulminant meningoencephalitis
-Diagnose by looking for trophozoites in the CSF or brain biopsy
-No cure
-Poor prognosis
- -Usually fatal

11. Olfactory mucosa
12. Mucor

Mucor and Rhizopus
-Invading fungi
-Common in diabetics
-Cause sinusitis which can spread rapidly to the CNS causing a cavernous sinus thrombosis and intracranial abscess
-Culture fluid that may be draining
-Histopathology shows nonseptated hyphae
-Treatment is emergent surgical debridement and IV amphotericin B
-Consider stat ophthalmology consult if infection involves the eye

13. Cavernous sinus Thrombosis
14. Non septate hyphae
15. Emergency surgical debridement and Amp B
16. Pneumocystis carinii

Pneumocystis carinii
-Fungus
-Commonly infects patients who are immunosuppressed, especially HIV patients with CD4 count less than 200
-Signs and symptoms include:
    -Fever
    -Nonproductive cough
-Leukocytosis
-Elevated LDH
-Chest x-ray shows diffuse hilar infiltrates
-Obtain arterial blood gas (ABG) if patient in respiratory distress
    -PaO2 is usually less than 70
    -If so, give corticosteroids, which has been shown to decrease mortality
-Conduct bronchoalveolar lavage (BAL) to obtain sample with organism
-Silver stain reveals organism
-Treat with Bactrim for 3 weeks
-Prophylaxis with Bactrim if CD4 count less than 200

17. Steroids
18. Bactrim
19. Rotavirus

Rotavirus
-Double stranded RNA virus
-Most common cause of watery diarrhea in children
-Fecal-oral transmission
-Common in daycare with sick contacts
-Self-limited
-Consider oral rehydration or IV fluids

20. Double stranded RNA virus—see above

Musculoskeletal

34-year-old female is involved in a motor vehicle crash. She was a restrained driver. She complains of right Hip and knee pain that is very focal. When she was extricated from the mangled car, her right knee was flexed up against the dashboard. On exam, she is able to move all muscle groups except Right proximal leg movement. Her vital signs are stable.

1) What is the likely diagnosis in this patient?
    a. Achilles Tendon rupture
    b. Medial Epicondylitis
    c. Lateral Epicondylitis
    d. Hip dislocation

2) Why must this pathology be diagnosed and treated urgently?
    a. Risk of avascular necrosis right femoral head
    b. Risk of septic meningitis
    c. Risk of femoral artery occlusion
    d. None of the above

45 year old female presents to neurology service with 5 year history of constant bilateral lower extremity episodes of numbness and burning pain. She had a history of Motor vehicle collision years ago resulting in severe low back pain. Her vital signs are stable. MRI of cervical, thoracic, and lumbar spine show degenerative changes but no severe stenosis of the canal or foramen. She states that hot and cold water cause severe pain in her legs.

3) What is the diagnosis?
    a. Reflex Sympathetic dystrophy
    b. Osteoarthritis
    c. Achilles tendinitis
    d. None of the above

4) What is initial treatment?
    a. Boots
    b. Cast
    c. Amputation foot
    d. NSAIDS

56-year-old female with a long history of medically controlled depression presents to the primary care doctor with a long history of severe and diffuse joint pains. She says that she has tried ibuprofen and other narcotics, but these have not touched her lately. On exam she has no focal finding other than multiple areas of tenderness to palpation throughout her musculature in her entire body. She weighs 245 pounds. Her labs are all normal.

5) What is the likely diagnosis?
    a. Fibromyalgia
    b. Early malaria
    c. Reiters syndrome
    d. Compartment syndrome

6) What portion of her past medical history likely contributes to her current pathology?
    a. obesity
    b. depression

c. Hypertension
d. Use of NSAIDS

7) What is the treatment of choice?
   a. Steroids
   b. Valproic acid
   c. PT and Antidepressants
   d. No treatment exists

8) Which of the following pathology might patients with this disorder also have?
   a. Irritable bowel syndrome
   b. Uterine cancer
   c. Inguinal Hernia
   d. Appendicitis

9) Posterior dislocations of the shoulder are at risk of injuring what artery and nerve?
   a. Axillary artery
   b. Radial artery
   c. Ulnar artery
   d. Brachial artery

10) A 45-year-old patient complains of knee pain. On exam he has a positive Lachman sign and Anterior drawer sign. What is the most likely pathology?
    a. PCL tear
    b. ML tear
    c. ACL tear
    d. None of the above

11) How can one make the diagnosis of knee dislocation?
    a. Knee xray
    b. MRI pelvis
    c. Ultrasound knee
    d. EMG/NCS

12) A 32-year-old male soccer player is hit on the medial aspect of his left leg. He complains of left knee pain. His vitals are stable. On, exam he has tenderness to palpation of the lateral aspect of the left knee. What other test should be positive in this patient?
    a. Varus test
    b. Valgus test
    c. Anterior drawer sign
    d. Posterior drawer sign

13) What physical exam test is indicative of an posterior cruciate ligament tear?
    a. Positive posterior drawer sign
    b. Positive Achilles Tap sign
    c. Positive anterior drawer sign
    d. None of the above

14) What injury is associated with a pop heard while walking with tenderness over medial knee and positive McMurray Sign?
    a. Medial Meniscal Injury
    b. Lateral Meniscal Injury
    c. Achilles tendinitis
    d. None of the above

15) What injury is associated with a pop heard while walking with tenderness over lateral knee and positive McMurray Sign?
    a. Medial Meniscal Injury
    b. Lateral Meniscal Injury
    c. Posterior dislocation
    d. Anterior dislocation

30-year-old male returned from a trip to Arizona where he battled a bout of diarrhea that last 2 days. He complains now of right eye pain and redness. He states he has joint pains all over which is something he has not experienced before. He states that "it burns when I pee."

16) What is the likely diagnosis?
    a. Ankylosing Spondylitis
    b. Osteoarthritis
    c. Lupus
    d. Reiter's Syndrome

17) What allele is often positive in these patients?
    a. HLA B 32
    b. HLA B 27
    c. HLA B 10
    d. None of the above

18) What diagnostic procedure should be done next?
    a. Aspirate septic joint
    b. CT Head
    c. Immediately start IV Antibiotics
    d. None of the above

19) What part of the septic joint fluid is crucial for the diagnosis in this pathology?
    a. Gram stain
    b. Culture
    c. Lymphocytes
    d. none of the above

Answers: Musculoskeletal

1) Hip dislocation

2) Risk of avascular necrosis of right femoral head
3) Reflex Sympathetic dystrophy

4) NSAIDS
5) Fibromyalgia
6) Depression
7) PT and anti-depressants
8) Irritable bowel syndrome
9) Radial artery

Posterior dislocations
    -Seen in seizures and electrocutions
    -Patients typically hold arm in internal rotation
    -Radial artery at risk

10) ACL tear

Anterior Cruciate Ligament (ACL) Tears
-May hear "pop"
-Common hyperextension injury
-Positive anterior drawer sign
-Positive Lachman test
    -More sensitive sign

11) Knee xray
12) Positive varus test

Lacteral Collateral Ligament (LCL) Tear
-Rare
-Caused by blow to the medial leg
-Tenderness on lateral side of knee when palpated
-Positive varus test

13) Positive posterior drawer sign

Posterior Cruciate Ligament (PCL) Tears
-Positive posterior drawer sign

14) Medial Meniscal Injury

Medial Meniscus Injury
-Hear classic "pop"
-Difficultly duck walking
-Tenderness over medial side of the knee
-Positive McMurray test

Lateral Meniscus Injury

-Hear classic "pop"
-Difficulty duck walking
-Tenderness over lateral side of the knee
-Positive McMurray

    15) Lateral Meniscal Injury
    16) Ankylosing Spondylitis

    17) HLA B27
Reiter's Syndrome
-Usually seen in young males
-HLA B27 positive
-Triad of conjunctivitis, urethritis, and arthritis (i.e., "cant see, pee, or climb a tree")
-Associated with gastrointestinal or genitourinary infection
        -Chlamydia most classically
        -Campylobacter jejuni
        -Salmonella
        -Shigella
-Negative rheumatoid factor
-Aspirate septic joint
        -Gram stain shows no organisms
            -Culture may reveal intracellular Chlamydia organism
            -Can do urine PCR or run PCR on joint aspirate to detect organism

    18) Aspirate septic joint
    19) Culture joint fluid

Rheumatology

A 30-year-old female gives birth to a healthy male. A few days later physicians diagnose a congenital heart block in this child. He is diagnosed with congenital lupus.
1) What antibody is typically seen in these patient's?
    a. Anti-Ro
    b. Anti-La
    c. Anti-Jo
    d. None of the above

A 45-year old-female has severely dry eyes and dry mouth. Her Vitals are stable. On exam, she has an enlarged region over the right side of her neck that is not tender to palpation. She has positive ANA and elevated ESR.
2) What HLA is usually positive in these patients?
    a. HLA DR 4/5
    b. HLA DR4
    c. HLA DR7
    d. HLA DR 3

3) What is treatment for this patient?
    a. Artificial tears and pilocarpine
    b. Steroids
    c. Antibiotics
    d. None of the above

4) What antibody is specific for SLE but is not present in discoid lupus?
    a. Anti-Ro
    b. Anti-So
    c. dsDNA
    d. ssDNA

5) 23-year-old woman is seen in the ED. She presents with fevers and a new rash on her cheeks. She also states that her urine is more foamy than normal. She has a history of hypertension well-controlled with hydralazine. She also takes Tylenol and ibuprofen. Her brother states that he has witnessed her having a seizure once a month. What is the most likely diagnosis?
    a. Sjogrens
    b. CREST Syndrome
    c. Scleroderma
    d. Drug-Induced SLE

6) What is the likely culprit that should be discontinued?
    a. Tylenol
    b. Hydralazine
    c. Ibuprofen
    d. Pepcid

7) What antibody is commonly seen in this disorder?
    a. Anti-Ro
    b. Anti-La

c. Anti-Histone
   d. Anti-Nuclear Antibody

8) What drugs can cause lupus?
   a. Hydralazine only
   b. Hydralazine, methyldopa and pepcid
   c. Methyldopa only
   d. Penicillamine, methyldopa, and quinidine

35-year-old obese female presents to ED with acute pain in right large toe and ankle joint. She recently returned from a party where she drank quite a bit of wine. Her vitals are stable. On physical exam, she has a large and reddened right big toe and ankle. She denies any recent trauma. She states that this has happened before.

9) What is the diagnosis?
   a. Sjogren's Disease
   b. Lupus
   c. Rheumatoid Disease
   d. Gout

10) Beside history and physical exam, what other methods of diagnosis can be done?
    a. Blood cultures
    b. CT Head
    c. MRI joint
    d. Aspirate fluid from affected joint

11) What HLA is usually positive?
    a. Rhomboid shaped crystals
    b. Needle shaped positively birefringent crystals
    c. Needle shaped negatively birefringent crystals
    d. Rhomboid shaped positively birefringent crystals

12) What is the typical first-line treatment?
    a. NSAIDS
    b. Aspirin
    c. Colchicine
    d. Steroids

13) Her family physician places her on chronic allopurinol. What is the mechanism of action for this drug?
    a. Increase ammonia production
    b. Decreased uric acid production
    c. Increase uric acid excretion
    d. None of the above

24-year-old female presents to ED stating that she noticed her hands turn white, then blue, then red. She has also had some difficulty swallowing solids. On exam, her fingers look very stiff and fibrotic, with tiny blood vessels on the skin of her body.

14) What is the diagnosis?
    a. CREST Syndrome

b. Diffuse Esophageal injury
c. Achalasia
d. Nutcracker Syndrome

15) What is the term for the changes in her hands?
    a. Vascular insufficiency Syndrome
    b. Loss of Autoregulatory syndrome
    c. Raynaud's Phenomenon
    d. Raymond's Phenomenon

16) What antibody is positive?
    a. Anti-centromere antibody
    b. Anti-Ro antibody
    c. FAB antibodies
    d. None of the above

17) What does the "T" in CREST stand for?
    a. Tumors
    b. Tremors
    c. Telangiectasia
    d. None of the above

35-year-old female presents to her family physician with muscle aches and a new rash on her shoulders. She has noticed that her eyelids look a little darker, as well as her knuckles. On exam she has 3/5 strength of bilateral hip flexors and extensors, and 5/5 strength in Bilateral dorsiflexors and plantar flexors. This pattern is seen in her upper extremities as well. Her sensation is intact to light touch and pin prick. Her reflexes are symmetric. She has normal rectal tone.

18) What is likely diagnosis?
    a. Polymyositis
    b. Dermatomyositis
    c. Freidrich's Ataxia
    d. Duchenne Muscular Dystrophy

19) Muscle biopsy is done, along with a nerve biopsy. Muscle biopsy shows what finding in this pathology?
    a. Perifascicular atrophy
    b. Significant collection of lymphocytes
    c. Excess fatty deposition in the muscle
    d. None of the above

20) What is the treatment of choice?
    a. NSAIDS
    b. Aspirin
    c. Prednisone
    d. None of the above

Answers: Rheumatology

1) ANTI-Ro antibody
2) HLA DR3
3) Artifical tears and pilocarpine

Sjogren's Syndrome (SS)
-Autoimmune disease
-HLA DR3 positive
-Seen in women in their 20s-30s
-Signs and symptoms include:
    -Dry eyes (i.e., no tears when crying)
    -Dry mouth (i.e., xerostomia)
    -Vaginal dryness
    -Parotid gland hypertrophy
-Positive anti-SSA (Ro), SSB (La)
-RF may be positive
-ANA may be positive
-Elevated ESR
-Treat with artifical tears, pilocarpine to increase tear production, NSAIDs for arthritis, and consider cyclosporine
-Complications include:
    -Lymphoma
    -Bacterial infection of dry eyes
        -Tears provide protection against organisms with lysozyme
    -Tooth cavities
        -Saliva protective

4) Anti-dsDNA
5) Drug-induced SLE
6) Hydralazine
7) Anti-Histone Antibody
8) Penicillamine, methyldopa, and quinidine

Drugs that cause lupus-like syndrome – patients will present with malar rash, but no renal or neurologic manifestations
-Hydralazine
-Procainamide
-Methyldopa
-Chlorpromazine
-Pencillamine
-Quinidine

9) Gout

Gout
-Seen in obese men
-Intense pain usually in the first large toe (i.e., first metatarsal, termed podagra
-Exacerbations and remissions
-Aggravated by red wine, meat, stress, and dehydration

- Secondary Gout
    - Lesch-Nyhan Syndrome
        - X-linked recessive
        - Defect in HGPRT enzyme
        - Self-mutilation
    - Tumor Lysis Syndrome
- Aspirate joint
    - Detect needle-shaped, negatively birefringent crystals
- Acute treatment – NSAIDs (i.e., indomethacin), colchicine (not first-line secondary to gastrointestinal side effects)
- Chronic prophylaxis – allopurionol (i.e., to decrease production), probenecid (i.e., to increase excretion)
- Do not give thiazides or furosemide

10) Aspirate fluid from affected joint
11) Needle-shaped negatively birefringent crystals
12) NSAIDS
13) Decrease production of uric acid
14) CREST Syndrome

CREST Syndrome
- Localized scleroderma
- Signs and symptoms include:
    - Calcinosis
    - Raynaud's phenomenon
        - Exacerbated by cold weather or stress
        - Vessel walls are fibrosed causing increased sensitivity to constriction
        - Hands turn white, blue, then red
            - White first because vessels vasoconstrict in response to cold
            - Turns blue due to lack of oxygen
            - Vessels then dilate causing flushing red
    - Esophageal dysfunction (i.e., GERD)
    - Sclerodactyly (i.e., fibrosed fingers)
    - Telangiectasias (i.e., dilated vessels)
- Positive anti-centromere antibody
- No cure exists
- Avoid cold or smoking for Raynaud's phenomenon and if severe use calcium channel blockers
- Consider antacids for GERD

15) Raynaud's Phenomenon
16) Anti-centromere antibody
17) Telangiectasia
18) Dermatomyositis

Dermatomyositis
- Signs and symptoms include:
    - Symmetric proximal muscle weakness
    - Shawl sign – rash on posterior neck and shoulders
    - Gottron's patches on knuckles
    - Heliotrope rash on eyelids
- Associated with interstitial lung disease and underlying malignancy

-Normal CK level
-Positive anti-Jo antibody
-Workup involves EMG, muscle biopsy, and looking for occult malignancy
-Muscle biopsy shows perifascicular atrophy
-Treat with prednisone

19) Perifascicular atrophy
20) Prednisone

Endocrinology

42-year-old male presents to the ED with headaches and weight gain. She has no visual field deficits. Her BP is 156/45. One exam, she is noted to have a large hump on the posterior aspect of her neck. She has a blood sugar of 350. She has significant abdominal obesity as well as multiple stretch marks throughout her body. Her face is very round with protuberant cheeks. What test should be done next?

1) What labs should be obtained?
    a. Full pituitary panel
    b. EKG
    c. CT chest
    d. MRI chest

2) CT head is done showing an area of hyperdensity within the sella turcica. What study should be ordered next?
    a. MRI Neck
    b. MRI Brain with pituitary cuts
    c. LP
    d. None of the above

3) What are some other causes of Cushing Syndrome?
    a. Prednisone
    b. Adrenal hyperplasia
    c. Ectopic ACTH tumor
    d. All of the above

4) A low dose dexamethasone suppression test shows no suppression. A high dose dexamethasone suppression test is done and this shows suppression. What is the likely cause?
    a. Ectopic tumor
    b. Pituitary tumor
    c. Cushing disease
    d. Both b and c

5) A 34-year-old patient is diagnosed with an ectopic ACTH-secreting tumor in the lung. She has noticed areas of hyperpigmented skin throughout her body. What is the likely cause of this?
    a. Tumor cells
    b. Alpha- MSH increase
    c. Melanocortin increase
    d. OCP use at home

6) A 4-year-old child is brought in by parents due to an increase swelling along the middle of his neck. He has no difficulty breathing but they are very concerned. Which of the following is the likely diagnosis?
    a. Branchial cyst
    b. Thyroglossal cyst
    c. Thyroid tumor
    d. Vagal nerve sheath schwanomma

7) What can be used next if bromocriptine does not work or if patient has allergy to it?
   a. pergolide
   b. Octreotide
   c. Sandostant
   d. Dopamine

34-year-old male has noticed that his shoes are too tight on him. He has noticed that his jaw has been enlarging. He has had shortness of breath as well of late with a recent diagnosis of Diabetes as well. His CT head showed a pituitary mass.

8) What is the diagnosis?
   a. Acromegaly
   b. Gigantism
   c. Bony Tumor
   d. None of the above

9) 45-year-old female presents to ED with weight loss and chest palpitations. She has noticed this over the last few months. Her pulse is 110. Her blood pressure is 124/68. On exam, she has on obvious swelling along the midline of her neck and has a bruit over that region. What other finding is commonly seen in these patient's?
   a. Cardiac tumors
   b. Exophthalmia
   c. Visual loss
   d. Spastic hemiparesis

10) What antibody is commonly seen in these patient's?
    a. Anti-Ro
    b. Anti-La
    c. Anti-TSH antibody
    d. Anti-T4 antibody

11) Radiaoctive iodide is used for treatment. What of the following should not have this treatment?
    a. Pregnant patients
    b. Patients on NSAIDS
    c. Patients with liver disease
    d. None of the above

12) What is the first-line treatment for pregnant women?
    a. Methimazole
    b. Iodide
    c. PTU
    d. NSAIDS

34-year-old female has noticed that she has abundant energy and rarely needs sleep. She has noticed that she can see more of her eyes now. Over the last few months however, she has noticed that her hair has become more dry and coarse, and her energy level is dramatically low. She has noticed some swelling in her legs recently.

13) What labs should be sent?

a. TSH, Free T3/4
   b. TSH only
   c. Reverse T3 only
   d. None of the above

14) TSH is high and T3/4 are low. What is the likely diagnosis?
   a. Hashimoto's Thyroiditis
   b. Follicular adenitis
   c. Graves disease
   d. MEN 2

15) What antibody is seen in this disorder?
   a. T3 antibody
   b. Anti-TSH antibody
   c. T4 antibody
   d. Anti-microsomal antibodies

16) What other antibodies may be seen in this pathology?
   a. Anti-Ro
   b. Anti-La
   c. Anti-histone
   d. Antibodies against peroxidase enzymes

17) If radioactive iodide is given to this patient, what is usually seen?
   a. Significant uptake in one area
   b. Significant uptake diffusely
   c. Minimal uptake in one area
   d. Minimal uptake diffusely

18) 23-year-old female presents with abdominal distention due to no bowel movement in 7 days. She also has been recently diagnosed with gout and peptic ulcer disease. She has been complaining of joint pains for the last year now. What is the likely diagnosis?
   a. Hyperparathyroidism
   b. Hyperthyroidism
   c. Thyroid cancer
   d. pseudohypoparathyroidism

19) What is the most common cause of this pathology?
   a. Adenoma
   b. Liver cancer
   c. Pituitary tumor
   d. Ectopic lung tumor

20) What value must be obtained to give an accurate serum calcium level?
   a. Total calcium
   b. Ionized calcium
   c. Serum PTH level
   d. Serum Growth hormone level

Answers: Endocrinology

1) Full pituitary panel

2) MRI Brain with pituitary cuts

3) All of the above
4) Both b and C
5) Alpha MSH increase
6) Thyroglossal cyst
7) Acromegaly

Acromegaly
-Disease caused by growth hormone (GH) pituitary tumor
-Secretion of GH occurs after fusion of epiphyses
-Signs and symptoms include:
    -Bitemporal hemianopsia
    -Headache
    -Coarsening facial features
        -Large jaw
        -Frontal bossing
        -Wide spaced front teeth
    -Enlarged hands
        -Require larger sized gloves
        -Ring does not fit on finger anymore
    -Enlarged feet
        -Require larger sized shoes
-Long bones grow width wise instead of length wise
-Complications include:
    -Carpal tunnel syndrome
    -Hyperglycemia
    -Type II diabetes mellitus
    -Hypertrophic cardiomyopathy
        -Most common cause of death
    -Increased risk of colon cancer
-Growth hormone antagonizes the effects of insulin
-Usually effects are secondary to insulin growth factor 1 (IGF-1) secreted by liver in response to GH
-Diagnosis involves measuring serum IGF-1 levels and conducting a growth hormone suppression test
    -Give load of glucose and measure GH levels, which should normally be low given negative feedback, but if elevated is suggestive of a GH secreting tumor
-Diagnosis also involves pituitary MRI to detect the tumor
-Treatment is octreotide
-Surgical resection indicated if refractory to octreotide
-Gigantism is similar to acromegaly except that GH secretion occurs before the epiphyses close
    -Long bones grow length wise instead of width wise
        -Very tall
        -Long limbs

8) Acromegaly

9) Exophthalmia
10) Anti-TSH antibody
11) Pregnant patient's
12) PTU
13) TSH, Free T3/4

Graves disease
-Autoimmune disease
-Type II hypersensitivity reaction
-Seen in women in their 20s-30s
-Signs and symptoms include:
- Diffuse goiter
- Pretibial edema
- Exophthalmos
- Bruit heard over thyroid
-Low TSH
-High T3 and T4
-Positive anti-TSH antibody which causes hyperthyroidism
-Elevated ESR
-If patient presents with hyperthyroid state with diffuse goiter, then after checking TFTs, test radioactive uptake to demonstrate areas of large uptake that is seen in Graves disease or can immediately start treatment after detecting antibodies consistent with Graves disease
-Thyroid biopsy which should not be done unless diagnosis unclear may reveal large thyroglobulin content within the center of the follicle
-Treat with radioactive iodide as first-line treatment in non-pregnant patients
- Complications include hypothyroidism
- Contraindicated in pregnancy and breastfeeding as can cause cretinism
-Second-line agents include propylthiouracil (PTU) and methimazole
- PTU first-line in pregnant women
- Methimazole contraindicated in pregnancy
- Both decrease synthesis of hormones, but PTU also inhibits conversion of T4 to T3
- Side effects of PTU includes agranulocytosis
-Pregnant women with Graves disease may have neonates who have hyperthyroidism secondary to antibodies (i.e., IgG) transferred via the placenta

14) Hashimoto's thyroiditis
15) Anti-microsomal antibodies
16) Antibodies against peroxidase enzymes
17) Minimal uptake diffusely
18) Hyperparathyroidism
19) Adenoma
20) Ionized calcium

Renal/Genitourinary System

34-year-old female has been battling a skin infection. She has been on penicillin for the last 2 weeks. She presents to the clinic and has a urine sample sent. Her BUN/Creat ratio is 10:1 with eosinophils seen in the urine.

1) What is the likely diagnosis?
    a. Allergic Interstitial Nephritis (AIN)
    b. Acute Tubular Necrosis
    c. Chronic Interstitial Nephritis
    d. None of the above

2) What is the cause of this pathology in our patient?
    a. Skin Infection
    b. Lack of IVF
    c. Penicillin use
    d. Unknown

3) What other finding on the urine sample confirms the diagnosis?
    a. White blood cell casts
    b. Red blood cell casts
    c. Neutrophils in Urine
    d. Opiates in Urine

4) What is treatment?
    a. NSAIDS
    b. Bactrim
    c. Stop penicillin
    d. None of the above

54-year-old male presents with acute onset of severe right low back pain. He says that pain goes from the back to his groin. He had this happen last year and the pain went away after 2 days. UA shows RBC and no casts.

5) What study should be done next?
    a. KUB
    b. CT head
    c. CT abdomen
    d. MRI abdomen

6) KUB will show all kidney stones except for what type?
    a. Uric acid and cystine stones
    b. Staghorn stones
    c. Gallstones
    d. Calcium oxalate stones

7) KUB does not show kidney stones. What study should be done next?
    a. LP
    b. Ultrasound abdomen
    c. CT Abdomen/pelvis

d. MRI Pelvis

8) CT abdomen shows multiple large stones in the right kidney. What is recommended treatment?

   a. Lithotripsy and IVF
   b. Radiation
   c. Intrathecal baclofen
   d. No treatment modality exists

9) If the patient had recurrent calcium oxalate stones, what syndrome must be ruled out?
   a. Hyperparathyroidism
   b. Addison Disease
   c. Conn Syndrome
   d. None of the above

10) What kind of stones are caused by proteus UTI?
    a. Magnesium Ammonium Phosphate Stones (i.e., struvite stones)
    b. Oxalate
    c. Uric acid
    d. Cystine

11) What is the treatment for struvite stones?
    a. Bactrim
    b. NSAIDS
    c. Morphine
    d. Chemotherapy

12) What pH is seen in the urine of patients with struvite stones?
    a. Acidic
    b. Basic
    c. Neutral
    d. Combination

13) What is the treatment for uric acid stones?
    a. Allopurinol
    b. Bromocriptine
    c. Octreotide
    d. Morphine

14) What disease must be ruled out in these patients?
    a. Leukemia tumor lysis syndrome
    b. Kidney Failure
    c. Liver Failure
    d. Sepsis

15) What medication must these patients avoid taking?
    a. Aspirin
    b. Beta Blockers
    c. Thiazides

d. Statins

16) What stones are seen in urine that is acidic?
    a. Cystine
    b. Oxalate
    c. Uric Acid
    d. None of the above

17) What is the cause of these stones developing?
    a. Congenital defect in reabsorption of the amino acid cystine from the urine
    b. Too much protein in diet
    c. Too little protein in diet
    d. It is not known

45-year-old female with a history of nephropathy. She also has hepatitis B and C, as well as lupus.

18) What does her kidney biopsy show?
    a. Spike and dome appearance secondary to deposit of IgG and C3 at the basement membrane
    b. Foot processes attached
    c. Pericytes absent
    d. None of the above

19) What is treatment?
    a. Heparin drip
    b. Chemotherapy
    c. NSAIDS
    d. Steroids

20) What type of nephropathy shows thickened mesangial matrix, Kimmelsteil-Wilson nodules (i.e., focal nodular glomerulosclerosis), or diffuse hyalinization (i.e., diffuse glomerulosclerosis)?
    a. Diabetic nephropathy
    b. Traumatic nephropathy
    c. Goodpasture's
    d. Unknown

Answers: Renal/Genitourinary System

1) Allergic Interstitial Nephritis (AIN)

-Allergic Interstitial Nephritis (AIN)
- -Usually secondary to drugs
  - -Penicillins
  - -Cephalosporins
  - -Furosemide
  - -Bactrim
- -BUN/Cr Ratio 10:1
- -Urine Na > 40
- -FeNa > 2 %
- -Eosinophils and white cell casts in the urine
- -Reversible with removal of drug

2) Penicillin use
3) Wbc casts
4) Stop penicillin
5) KUB

Kidney Stones
-Patients present with abdominal pain on either side that radiates to their groin
-Some stones may be asymptomatic
-Can be severely painful
- -Patient usually writhing in pain and moving around
- -Contrast to patients with peritonitis where patients are lying still
-Urinalysis reveals RBCs and no casts
-Kidney-Ureter-Bladder (KUB) x-ray may show presence of all stones except uric acid and cystine stones
-Renal ultrasound may show hydronephrosis
-CT abdomen/pelvis with contrast (i.e., CT Urogram) can detect all types of stones
-Treatment is hydration and analgesia
-Prazosin or terazosin can be given to flush out the stone
-Procedures for recurrent stones include lithotripsy

6) Uric acid and cystine stones

Four Types of Kidney Stones
-Calcium Oxalate Stones
- -Most common type of stone
- -Causes include:
  - -Hyperparathyroidism
  - -Hypercalcemia (i.e., and its causes)
  - -Malabsorption syndromes (i.e., post-ileal resection)
-Calcium binds oxalate in the urine and allows us to excrete oxalate, so if the patient takes in low diet of calcium or has any state of malabsorption, oxalate can precipitate in the urine
- -Urinalysis may reveal oxalate crystals
-Ask patient to bring in stone for analysis

- Usually first do a 24 hour urine collection to measure urine calcium to detect hypercalciuria
- Can check a parathyroid hormone (PTH) level in patients if you suspect hyperparathyroidism
- Treatment is hydration and analgesia
- Thiazides may reduce urinary calcium
- Treat underlying cause
- Magnesium Ammonium Phosphate Stones (i.e., struvite stones)
  - Called staghorn calculi if large enough to obstruct the kidney-ureter junction
  - Caused by Proteus urinary tract infection (UTI), which produces urease that breaks down urea to ammonia and carbon dioxide
  - Urinalysis reveals basic pH secondary to ammonia
  - Urine culture helpful to detect bacteria
  - Treat UTI with Bactrim or ciprofloxacin
- Uric Acid Stones
  - Caused by elevated serum uric acid
  - Seen in leukemia secondary to tumor lysis syndrome, gout, or Lesch-Nyhan syndrome
  - Do not give thiazides to these patients as they cause hyperuricemia
  - Urinalysis shows acidic pH (i.e., less than 5.5)
  - These stones are not visualized on a KUB
  - Treat with allopurinol (i.e., decrease production of uric acid by inhibiting xanthine oxidase)
- Cystine Stones
  - Rare
  - Caused by congenital defect in reabsorption of the amino acid cystine from the urine
  - Urinalysis shows acidic pH (i.e., less than 5.5)
  - These stones are not visualized on a KUB

7) CT Abdomen/pelvis
8) Lithotrypsy, IVF
9) Hyperparathyroidism
10) Magnesium Ammonium Phosphate Stones (i.e., struvite stones)
11) bactrim
12) basic
13) allopurinol
14) Leukemia tumor lysis syndrome
15) thiazides
16) cystine
17) congenital defect in reabsorption of the amino acid cystine from the urine
18) spike and dome appearance secondary to deposit of IgG and C3 at the basement membrane

Membranous Glomerulonephritis
- Most common cause of nephropathy in adults
- Associated with hepatitis B and C, penicillamine, and lupus
- Kidney biopsy reveals spike and dome appearance secondary to deposit of IgG and C3 at the basement membrane
- Treat with corticosteroids
  - Usually not responsive
  - Progressive disease

19) steroids
20) Diabetic nephropathy

Diabetic Nephropathy
-History of diabetes
-Kidney biopsy reveals thickened mesangial matrix, Kimmelsteil-Wilson nodules (i.e., focal nodular glomerulosclerosis), or diffuse hyalinization (i.e., diffuse glomerulosclerosis)
-Treatment is to control glucose and prescribe angiotensin converting enzyme (ACE) inhibitors to decrease progression of proteinuria

Electrolytes and Acid-Base Disturbances

34-year-old male presents to ED with low back pain and constipation. He states all his bones hurt as well. He has a past medical history significant for multiple fractures of his arms and legs.

1) What lab should be sent?
    a. Calcium level
    b. Thyroid level
    c. PTT
    d. INR

2) Calcium level returns at 13.5. What study should be done next?
    a. CT head
    b. MRI Head
    c. 12 lead EKG
    d. Ultrasound neck

3) What are two causes for hypercalcemia?
    a. Pituitary Tumors and Zollinger Ellison
    b. Hyperparathyroidism and zollinger-Ellison Syndrome
    c. Hyperparathyroidism and liver failure
    d. Liver Failure and Kidney Failure

4) Why is the patient having low back pain?
    a. Kidney stones
    b. Gallstones
    c. Lumbar spondylosis
    d. None of the above

5) What is treatment?
    a. IVF and lasix
    b. Dilaudid
    c. Steroids
    d. Immediate surgery

23-year-old female presents to clinic with occasional episodes of numbness around her lips along with weakness of her arms and legs. On exam, she has facial contraction when her facial nerve is tapped.

6) What other exam sign might she have?
    a. Chvostek Sign
    b. Tremor Sign
    c. Trousseau sign
    d. None of the above

7) What lab should be sent?
    a. Albumin
    b. INR
    c. Ionized Magnesium
    d. Ionized calcium

8) What is the diagnosis?
    a. Hypocalcemia
    b. Hypermagnesemia
    c. Elevated Protein in serum
    d. Vasovagal

9) What is treatment?
    a. Po calcium with magnesium and Vitamin D
    b. IV Synthroid
    c. Phosphate
    d. Magnesium only

10) What is the most common cause of hypocalcemia?
    a. Hyperparathyroidism
    b. Hypertension
    c. Hypothyroidism
    d. Hypoparathyroidism

11) Aspirin overdose causes what changes on ABG?
    a. First respiratory alkalosis, then metabolic acidosis
    b. First Respiratory alkalosis, then metabolic alkalosis
    c. First Metabolic alkalosis, then respiratory acidosis
    d. None of the above

12) TCA use must be monitored by what test?
    a. CT chest
    b. MRI brain
    c. EKG
    d. Calcium levels

A 30 kg male is being admitted for dehydration.

13) What rate should his IVF be running at for maintenance?
    a. 60 cc/hour
    b. 70 cc/hour
    c. 75 cc/hour
    d. 80 cc/hour

A 45-year-old Alcoholic male presents with numbness and tingling in his hands and feet. He also states that he has occasional shortness of breath when he runs. On exam, he has a right lateral rectus palsy with gait instability.

14) What is the likely diagnosis?
    a. Wernicke Encephalopathy
    b. Korsakoff Syndrome
    c. Ataxia telangiectasia
    d. Dry beriberi

15) What is the cause of this syndrome?

a. B12 deficiency
b. Niacin deficiency
c. Vitamin B1 deficiency
d. Riboflavin deficiency

16) What structure on CT or MRI is often seen to be small or with hemorrhage?
    a. Mamillary bodies
    b. Thalamus
    c. Internal Capsule
    d. Caudate nucleus

17) What is treatment?
    a. IV thiamine
    b. Riboflavin
    c. Niacin
    d. Accutane

18) Why might he have shortness of breath?
    a. Myocardial infarction
    b. Dry Beriberi cardiomyopathy
    c. Wet beriberi dilated cardiomyopathy
    d. Pulmonary embolus

19) A 67-year-old male has skin lesions, a 3 week history of diarrhea, and has been forgetful according to family. What vitamin is likely deficient?
    a. Niacin
    b. B12
    c. B5
    d. B3

20) A 23-year-old female has had difficulty seeing in the dark when she is driving. She also has very dry skin. What vitamin is she deficient in?
    a. Vitamin A
    b. Vitamin C
    c. Vitamin E
    d. None of the above

Answers: Electrolytes and Acid-Base Disturbances

1. calcium level

Hypercalcemia
-Elevated serum calcium level (i.e., greater than 10.5)
-Causes include ("CHIMPANZEES"):
    -Calcium supplementation
    -Hyperparathyroidism
        -Most common cause of hypercalcemia as outpatient is primary hyperparathyroidism
            -Most commonly caused by parathyroid adenoma
    -Milk-alkali syndrome
    -Paget's disease
    -Thiazides
    -Immobilization
    -Zollinger-Ellison syndrome
    -Acromegaly
    -Neoplasms
        -Most common cause of hypercalcemia as inpatient is malignancy (i.e., breast, prostate, lung, or multiple myeloma)
    -Vitamin A intoxication
    -Vitamin D intoxication
    -Sarcoidosis
-Signs and symptoms include (i.e., "stones, bones, moans, groans, and psychiatric overtones"):
    -Kidney stones
    -Bone pain
    -Fractures
    -Constipation
    -Nausea
    -Vomiting
    -Polyuria
    -Psychosis
-Diagnose with serum calcium level
    -Also check albumin level
-Conduct EKG
    -Note shortened QT interval
-Treat with IV fluids and furosemide
    -Do not give thiazides as they cause reabsorption of calcium from the tubules

2. 12 lead EKG
3. Hyperparathyroidism and zollinger-Ellison Syndrome
4. Kidney stones
5. IVF and Lasix
6. Trousseau sign

7. Ionized calcium
8. Hypocalcemia

Hypocalcemia
-Low serum calcium level (i.e., less than 8.5)
-Causes include:
    -Hypoparathyroidism
        -Most common cause
    -Acute pancreatitis
    -Hypomagnesemia
        -Must replete magnesium to correct calcium deficit
    -Vitamin D deficiency
-Signs and symptoms include:
    -Perioral numbness
    -Muscle weakness
    -Chvostek's sign – tapping of the facial nerve causes contraction of facial muscles
    -Trousseau's sign – inflation of blood pressure cuff causes carpal spasm
-Diagnose with serum calcium and phosphate levels
    -Do not forget to check albumin level and correct if necessary
    -Do not need albumin check if send ionized calcium level
    -Also check serum magnesium level
-Conduct EKG
    -Note prolonged QT interval
-Can give IV calcium over short-term if level severely low
-Treat with oral calcium and vitamin D over long-term
-Replete magnesium

9. Po calcium with magnesium and Vitamin D
10. Hypoparathyroidism
11. First respiratory alkalosis, then metabolic acidosis
12. EKG
13. 70 cc/hour

Calculation of IV Maintenance Fluids
-4-2-1 rule
    -4 ml/kg x first 10 kg
    -2 ml/kg x second 10 kg
    -1 ml/kg x any kg after 20 kg
    -Give that much per hour

14. Wernicke Encephalopathy

Thiamine B1 Deficiency
-Common in alcoholics
-Dry beriberi – peripheral neuropathy
-Wet beriberi – dilated cardiomyopathy and high output heart failure
-Cause of Wernicke-Korsakoff syndrome
    -Hemorrhage of the mamillary bodies of the hypothalamus
    -Wernicke's encephalopathy
        -Ophthalmoplegia (i.e., lateral rectus palsy)
        -Gait instability

                -Altered mental status
                -Reversible
       -Korsakoff's psychosis
                -Confabulation
                -Amnesia
                -Irreversible
-Patient courses from Wernicke's to Korsakoff's if left untreated
-Give IV thiamine in emergency room
       -Do not give glucose first as it may worsen thiamine deficiency and put patient into Korsakoff's psychosis
                -Glucose metabolism utilizes thiamine
-Also administer banana bag that is full of B vitamins

15. Vitamin B1 deficiency
16. Mamillary bodies
17. IV thiamine
18. Wet beriberi dilated cardiomyopathy
19. Niacin

Niacin B3 Deficiency
-Pellagra:
       -Diarrhea
       -Dementia
       -Dermatitis

20. Vitamin A

Vitamin A Deficiency
-Dry skin
-Night blindness

Obstetrics

1) What physiological changes occur during pregnancy:
    a. Tachycardia, increased stroke volume, increased cardiac output, increased Tidal volume
    b. Tachycardia only
    c. Increased stroke volume and tachycardia only
    d. Increased tidal volume and tachycardia only

2) What physiological changes are unchanged in pregnancy?
    a. Respiratory rate
    b. Pulse
    c. Stroke volume
    d. Blood Pressure

3) What happens to hematocrit during pregnancy?
    a. Increases
    b. Decreases
    c. Stays the same

4) What is the reason for hypothyroidism seen during pregnancy?
    a. Increased thyroid-binding globulin
    b. Increased Dopamine
    c. Decreased T3 made
    d. Fetus eats the thyroid hormone

5) What vitamin is recommended prenatally to prevent neural tube defects
    a. Niacin
    b. Folic acid
    c. Riboflavin
    d. B8

29-year-old female in second trimester of pregnancy presents with continued vomiting episodes that she has had since first trimester. What is the best test for diagnosis of this condition?
    a. Serum PTT
    b. Urine and serum ketones
    c. Serum calcium
    d. Urine Toxicology screen

    e. What is likely diagnosis?
        a. Hyperemesis gravidarum
        b. Nausea
        c. Food intolerance
        d. Abdominal pressure on the stomach

f. What is treatment?
   a. Larger infrequent meals
   b. IVF, small frequent meals
   c. Metoprolol
   d. Clonidine

42-year-old female recently found out that she is pregnant.

8) What procedure should be done to test for fetal anomalies?
   a. CT mother's Abdomen
   b. Amniocentesis
   c. MRI abdomen
   d. None of the above

9) Between what time frame can an amnio be done?
   a. Weeks 15-17
   b. Weeks 20-25
   c. Weeks 25-30
   d. Weeks 1-5

10) What can amniocentesis help with?
    a. Diagnosis sex
    b. Diagnose brain tumor
    c. Identify fetal lung maturity, rule out downs, evaluate RH incompatibility
    d. No information is gained from amniocentesis

11) When can Chorionic villous sampling be done?
    a. Weeks 10-12
    b. Weeks 15-20
    c. Weeks 20-25
    d. Weeks 1-5

35-year-old pregnant female has an elevated AFP level in serum.

12) What are some causes for this?
    a. Anencephaly only
    b. Anencephaly and spina bifida only
    c. Anencephaly, spina bifida, gastoschisis, omphalocele, multiple gestations
    d. None of the above

13) What is the most common cause of elevated Serum AFP?
    a. Incorrect gestation dating
    b. Brain tumor

    c. Kidney tumor
    d. Gestational Diabetes

14) What if her AFP level was low. What should one always rule out with amnio?
    a. Anencephaly
    b. Downs
    c. Megaencephaly
    d. Porencephaly

Mary is in Labor. She is 4 cm dilated at the cervix.

15) What state of labor is she in?
    a. First
    b. Second
    c. Third
    d. Fourth

16) What factors can prolong this first stage?
    a. Cephalopelvic Disproportion(CPD) only
    b. Increased contractions only
    c. Macrosomia and CPD only
    d. CPD, decreased contraction intensity, macrosomia

17) She has now delivered the fetus and is about to deliver the placenta. What stage is she in?
    a. Third
    b. First
    c. Second
    d. Fourth

She had developed diabetes during her second trimester. She has no history of diabetes. Her child was at the 90 % in weight.

18) What is the likely diagnosis?
    a. Congenital Diabetes
    b. Secondary to diet high in sugars
    c. Gestational diabetes mellitus
    d. None of the above

19) What is this associated with?
    a. Polyhydramnios and sacral agenesis
    b. Oligohydramnios
    c. Sacral agenesis only
    d. Anencephaly only

20) How can this be diagnosed in utero?
    a. 1 hour glucose challenge test at 24 weeks gestation
    b. 10 hour glucose challenge at 24 weeks
    c. Measuring Growth hormone and IGF-1 Levels
    d. Cannot be diagnosis in utero

21) What is treatment of choice?
    a. Diet management, insulin injections
    b. Diet management alone
    c. Pancreatectomy
    d. None of the above

32 female in her second trimester has an sbp of 175/120.

22) What must be checked next?
    a. UA
    b. Urine Culture
    c. CT head
    d. Ultrasound kidneys

23) What is this condition called?
    a. Pre-eclampsia
    b. Eclampsia
    c. HELLP syndrome
    d. None of the above

24) She later develops anemia, bruising, bleeding, elevated liver enzymes, and thrombocytopenia. What is treatment of choice?
    a. Induce with oxytocin only
    b. Induce with oxytocin for delivery, magnesium sulfate, IV labetolol
    c. Magnesium sulfate only
    d. Labetolol and magnesium sulfate only

25) What is magnesium sulfate used for?
    a. Nutrition
    b. Prevent Bleeding
    c. Prevent diabetes
    d. Prophylaxis against seizures

A 23-year-old woman in her last trimester states that she felt a gush of clear fluid from her vagina. She is not yet in her first stage of labor.

26) What test can be done to confirm this?
    a. Positive nitrazine blue test
    b. Negative nitrazine blue test
    c. None of the above

27) What is often the cause of this?
    a. Infection
    b. Trauma
    c. Liver failure
    d. Foley placement

28) What is treatment?
    a. NSAIDS
    b. Heparin
    c. IV antibiotics
    d. Steroids

29) Lithium is associated with what anomaly during pregnancy?
    a. Ebsteins anomaly
    b. Pulmonary fibrosis
    c. Renal failure
    d. Myocardial infarction

30) Dilantin is associated with what anomaly during pregnancy?
    a. Kidney failure
    b. Liver failure
    c. Mental retardation (MR), Dysmorphic craniofacial features
    d. Ebstein's Anomaly

31) DES is associated with what anomaly during pregnancy?
    a. Clear cell vaginal adenocarcinoma, vaginal adenosis
    b. Renal cancer
    c. Thyroid cancer
    d. None of the above

Answers: Obstetrics

1. Tachycardia, increased stroke volume, increased cardiac output, increased Tidal volume

Physiological Changes of Pregnancy
-Increased heart rate (HR)
-Increased stroke volume (SV)
-Increased cardiac output (CO)
-Increased systemic vascular resistance (SVR)
-Increased tidal volume (TV)
-Unchanged respiratory rate (RR)
-Increased minute ventilation (MV)
    -Cause of respiratory alkalosis
-Decreased hematocrit (Hct)
    -Physiological anemia of pregnancy
    -Secondary to increased plasma volume
-Hyperglycemia
    -Secondary to insulin resistance
-Increased thyroid hormone (i.e., T3 and T4)
-Increased thyroid-binding globulin
-Folic acid deficiency
    -Supplement prenatally to prevent neural tube defects
-Iron deficiency
    -Supplement during pregnancy
-Increased glomerular filtration rate (GFR)
-Increased acid reflux
-Chadwick's sign
    -Violet colorations of the vagina from increased blood flow
-Bloody show
    -Thick mucous plug in the cervical os that when expelled signals the onset of labor
-Hypercoagulable state
    -Increased risk of deep vein thrombosis (DVT)
-Increased weight
    -Gain of 25-35 lbs represents adequate weight gain during pregnancy in patients with normal body mass indices (BMIs)
-Hyperpigmentation
    -Linea nigra – seen over the abdomen
    -Melasma – seen on face
        -Secondary to estrogen
-Spider angiomas and palmar erythema
    -Secondary to estrogen
-Diastasis recti
    -Abdominal rectus muscles separate along the midline
-Morning sickness
    -Nausea and vomiting usually limited to the first trimester

2. respiratory rate
3. decreases
4. Increased thyroid-binding globulin

5. Urine and serum ketones
6. Hyperemesis gravidarum

Hyperemesis Gravidarum
-Pathologic state
    -Contrast with morning sickness
        -Usually morning sickness common within first trimester up until 12 weeks
-Vomiting that continues past the first trimester into the second trimester and begins to cause metabolic and electrolyte disturbances
-Hypokalemia
-Hyperchloremia
-Metabolic alkalosis with anion gap metabolic acidosis
-Positive urine and serum ketones
    -Best test for diagnosis
-Treat with increased fluids (i.e., remain hydrated) and small, frequent meals

7. IVF, small frequent meals—see above
8. Amniocentesis

Amniocentesis
-Samples amniotic fluid
-Can be performed at weeks 15-17
-Indications include:
    -35 years of age or older at time of delivery
    -To evaluate fetal lung maturity by examining the lecithin to sphingomyelin ratio
        -Greater than 2.5 suggests maturity
    -To rule out Down's syndrome if being considered
    -To evaluate Rh incompatibility in women at risk

Chorionic Villus Sampling (CVS)
-Samples villous tissue
-Can be performed at weeks 10-12
-Carries higher risk of adverse effects than amniocentesis

9. weeks 15-17
10. Identify fetal lung maturity, rule out downs, evaluate RH incompatibility
11. weeks 10-12
12. Anencephaly, spina bifida, gastoschisis, omphalocele, multiple gestations

α-Fetoprotein (AFP) Levels
-Elevated in the serum in:
    -Open neural tube defects
        -Anencephaly
        -Spina bifida
    -Gastroschisis
    -Omphalocele
    -Multiple gestations

  -Incorrect gestational dating
    -Most common cause
-Decreased in the serum in Down's syndrome
  -Must do amniocentesis and karyotyping

Quad Screen
-Down's syndrome
  -Low AFP
  -Low estriol
  -High β-hCG
  -High inhibin
-Trisomy 18
  -Low AFP
  -Low estriol
  -Low β-hCG
  -Low inhibin
-Trisomy 13
  -Normal AFP
  -Low β-hCG

13. Incorrect gestation dating
14. Downs
15. First
16. CPD, decreased contraction intensity, macrosomia
17. Third
18. Gestational diabetes mellitus

Gestational Diabetes Mellitus (GDM)
-Hyperglycemia in first trimester suggests preexisting diabetes mellitus
-Hyperglycemia developing without prior history of diabetes mellitus suggests gestational diabetes mellitus
-Associated with:
  -Large for gestational age (LGA) infants
    -Greater than 90th percentile in weight
    -Macrosomia defined as greater than 4500 grams
  -Polyhydramnios
    -Secondary to osmotic effect of glucose
  -Congenital defects
    -Sacral agenesis
    -Congenital heart defects
-The elevated glucose levels in the mother's blood crosses the placenta and causes the fetal pancreas to secrete insulin
  -Mother's insulin does not cross the placenta
  -Beware of hypoglycemia in the newborn after delivery secondary to elevated insulin levels
    -Elevated glucose levels of mother no longer entering fetal circulation
-Diagnosis in the mother involves first a 1 hour glucose challenge test at 24 weeks gestation
  -All women at high risk are screened including those who are African American, Native American, obese, and have other comorbidities

    -If serum glucose greater than 140, confirm with a 3 hour glucose challenge test
-Treatment depends on severity of gestational diabetes mellitus
    -May start with diet management and follow with insulin injections if dietary control insufficient
    -Oral anti-hyperglycemics are contraindicated during pregnancy
-Mothers with gestational diabetes are at increased risk of developing type II diabetes mellitus later in life

19. Polyhydramnios and sacral agenesis
20. 1 hour glucose challenge test at 24 weeks gestation
21. Diet management, Insulin injections
22. UA for proteinuria
23. Pre-eclampsia

Pre-Eclampsia and Eclampsia
-Usually develops in late second, early third trimester
-Rare in the first trimester
    -If pre-eclampsia signs and symptoms occur in first trimester, consider molar pregnancy
-Can be mild versus severe
    -Mild
        -Greater than 140/90 blood pressure
        -Proteinuria 500 mg/24 hours
        -No signs of end organ damage
    -Severe
        -Greater than 160/110 blood pressure
        -Proteinuria 5 g/24 hours
        -Signs of end-organ damage
            -Headache
            -Blurry vision
            -Right upper quadrant or epigastric pain
            -Elevated liver enzymes
            -Thrombocytopenia (i.e., less than 100,000)
            -Hemolysis
            -Beware of HELLP syndrome:
                -Hemolysis
                -Elevated liver enzymes
                -Low platelets or thrombocytopenia
        -Oliguria (i.e., less than 500 ccs/24 hours)
        -Liver capsule distention
            -May lead to a liver capsule hematoma
            -If hematoma ruptures, may be a cause for right upper quadrant pain in patient with pre-eclampsia
-Diagnose pre-eclampsia with history, blood pressure, and urinalysis or urine dipstick
-For mild pre-eclampsia, if at term, induce with oxytocin for delivery
    -Treat hypertension with IV labetalol or hydralazine
-For mild pre-eclampsia, if at preterm, consider expectant management
    -Give patient a dose of betamethasone for lung maturation in case of preterm delivery
    -Treat hypertension with IV labetalol or hydralazine
-For severe pre-eclampsia at any gestational age, induce with oxytocin for delivery
    -Give patient magnesium sulfate (MgSO4) for seizure prophylaxis and treatment

-Treat hypertension with IV labetalol or hydralazine
-Eclampsia defined as seizures in a patient with pre-eclampsia without any other known cause

24. Induce with oxytocin for delivery, magnesium sulfate, IV labetolol
25. Prophylaxis against Seizures
26. Positive nitrazine blue test

Premature Rupture of Membranes (PROM)
-Rupture of membranes at least 1 hour before the onset of the first stage of labor
-Usually secondary to infection
-Increases risk of chorioamnionitis and preterm labor
-Complaint of gush of clear fluid from vagina
- Positive nitrazine blue test
    -Turns blue secondary to alkaline fluid present
    -Can have false positive nitrazine blue test in bacterial vaginosis (BV) and Trichomonas infection
        -Workup also includes urinalysis and wet mount
-Positive fern test
    -Note fern pattern under microscope
-Do not perform digital vaginal exams in patients with PROM because of risk of infection spread
-Treatment is to give IV antibiotics (i.e., ampicillin and gentamicin or clindamycin and gentamicin) and induce labor with oxytocin if at greater than 34 weeks
    -If less than 32 weeks, start IV antibiotics and give dose of betamethasone before inducing labor with oxytocin

Preterm Premature Rupture of Membranes (PPROM)
-Rupture of membranes that occurs at less than 37 weeks of gestation
-Increases risk for chorioamnionitis
-Same management as above

27. Infection
28. IV antibiotics-AMP and GENT
29. Ebstein's Anomaly

Teratogens

Diethylstil (DES) besterol
-Causes:
    -Clear cell vaginal adenocarcinoma
    -T-shaped uterus
    -Vaginal adenosis

Lithium
-Causes Ebstein's anomaly

Phenytoin
-Causes fetal hydantoin syndrome
    -Mental retardation (MR)

-Dysmorphic craniofacial features

Valproic Acid
-Causes neural tube defects

Thalidomide
-Causes reduced limb growth

Vitamin A
-Causes microophthalmia

ACE Inhibitors
-Causes renal tubular dysplasia

Warfarin
-Causes nasal hypoplasia

Fluoroquinolones
-Causes tendon ruptures

Tetacyclines
-Causes:
    -Enamel discoloration
    -Limb shortening

Aminoglycosides
-Causes hearing loss

Cocaine
-Causes:
    -Spontaneous abortion
    -Placenta abruptio

Nicotine
-Causes low birth weight

Alcohol
-Causes fetal alcohol syndrome (FAS)
    -Mental retardation (MR)
    -Thin philtrum of the lip
    -Ventricular septal defects (VSDs)

30. Mental retardation (MR), Dysmorphic craniofacial features
31. Clear cell vaginal adenocarcinoma, vaginal adenosis

Gynecology

1) What structure degenerates 8 weeks after fertilization that maintained the pregnancy until the placenta takes over at 8 weeks?
   a. Corpus luteum
   b. Placenta
   c. Endometrium
   d. None of the above

24-year-old pregnant female was noted to have a mass on pelvic ultrasound on the right side. It is a fluid filled structure that is small.

2) What is the likely diagnosis at 8 weeks gestation?
   a. Endometrioma
   b. Corpus luteal cyst
   c. Leiomyoma
   d. Leiomyosarcoma

3) What is treatment?
   a. None required
   b. OR immediately
   c. Steroids
   d. Radiation therapy

Stephanie is a 34-year-old female who has constant pain each month during her menstrual cycle. She has been diagnosed with adenomyosis.

4) What is the term given to this pain syndrome?
   a. Secondary dysmenorrhea
   b. Primary dysmenorrhea
   c. Endometriosis
   d. Leiomyoma

5) What is treatment?
   a. Radiation
   b. Surgery
   c. NSAIDS, combined OCP
   d. Palliative care

6) If medical management dose not control her pain caused by adenomyosis, what is next treatment option?
   a. Radiation
   b. Surgical management
   c. Steroids
   d. Chemotherapy

44-year-old female with history of fibroids that have been stable in size though very painful.

7) What is the common initial treatment option?
    a. NSAIDS
    b. Heparin
    c. Chemotherapy
    d. None of the above

8) What other medication can be tried to decrease fibroid size?
    a. Leuprolide
    b. Metoclopramide
    c. Montelukast
    d. None of the above

9) If medical management does not decrease her pain and she continues to have bleeding and infertility, and desires to have kids, what treatment should be offered?
    a. Radiation
    b. Chemotherapy
    c. Intra fibroid injection steroids
    d. Myomectomy

10) What may fibroids develop into if ultrasound shows mass that has grown quickly over 2 months?
    a. Leiomyosarcoma
    b. Leiomyoma
    c. Chondrosarcoma
    d. Osteoleiomyosarcoma

24-year-old female with fevers, abdominal tenderness, and cervical motion tenderness. She has used IUD for many years.

11) What tests should be done next?
    a. CVC only
    b. CVC, culture vaginal discharge, and pelvic ultrasound
    c. Culture vaginal discharge and pelvic ultrasound
    d. None of the above should be done

12) What is often seen in patients with this condition on vaginal culture?
    a. Polymicrobial
    b. Streptococcus
    c. Staphylococcus
    d. Plasmodium species

13) What are the most common organisms causing this condition?
    a. Neisseria gonorrhea and Chlamydia trachomatis
    b. Steptococcus
    c. Staphylococcus
    d. None of the above

14) What is treatment of choice?
    a. IV or PO abx
    b. Intrathecal Abx
    c. Radiation
    d. No treatment exists

15) Why must an ultrasound be done?
    a. Rule out tubo-ovarian abscess
    b. Rule out hemorrhage
    c. Rule out large solid mass
    d. None of the above

16) PID caused by neisseria may lead to what syndrome?
    a. Fitz-Hugh Curtis syndrome
    b. Abdominal compartment syndrome
    c. MEN syndrome 2
    d. None of the above

33-year-old female who has endometritis and prior Dilatation and curettage procedures presents because she has not had a period for 6 straight months.

17) What syndrome is associated with this secondary amenorrhea?
    a. Asherman syndrome
    b. Endometrioma syndrome
    c. Cervical dislodgement syndrome
    d. Cervical hemorrhage

32-year-old obese female with hirsutism and scaly skin patches presents with headaches. Her blood glucose in 210.

18) What test must be done to attain the diagnosis?
    a. Pelvic ultrasound
    b. CT Head
    c. Ultrasound chest
    d. EKG

19) What is seen on pelvic ultrasound?
    a. Nutcracker ovaries
    b. Ovarian hemorrhages

    c. Ovarian cysts bilaterally (i.e., pearl necklace sign)
    d. None of the above

20) What hormones are often elevated?
    a. Testosterone only
    b. Estrogen only
    c. Estrogen, testosterone
    d. Estradiol only

21) What is treatment?
    a. OCP
    b. NSAIDS
    c. Immediate surgery
    d. Radiation therapy

22) What is treatment if she wants to get pregnant?
    a. Aspirin
    b. OCP
    c. Bisphosphonates
    d. Clomiphene

23) What side effect can occur with clomiphene use?
    a. Multiple gestations
    b. Kidney disease
    c. Pituitary tumor
    d. None of the above

Answers: Gynecology

1. Corpus Luteum

Corpus Luteal Cyst
-The corpus luteum degenerates after about 8 weeks after fertilization
    -It secretes estrogen and progesterone that maintains the pregnancy until the placenta begins to take over at 8 weeks of gestation
    -Requires β-hCG secreted by the placenta to remain active
-Cyst may be visualized via pelvic ultrasound in pregnant women within 8 weeks of gestation
-Also may be visualized in women who are not pregnant 2 weeks after their menstrual period
-Often unilateral in both cases
-Diagnose with clinical history and pelvic ultrasound revealing cyst
-No treatment required
-Bilateral corpus luteal cysts seen in molar pregnancies secondary to highly elevated β-hCG levels

2. Corpus luteal cyst—see above
3. None required
4. Secondary Dysmenorrhea

Primary Dysmenorrhea
-Menstrual pain during cycle
-Most often secondary to prostaglandins
-Treatment is NSAIDs
-If severe, try oral contraceptive pills (OCPs)

Secondary Dysmenorrhea
-Menstrual pain during the cycle with organic cause
-Causes include:
    -Endometriosis
    -Adenomyosis
    -Fibroids
    -Pelvic inflammatory disease (PID)
-Treat underlying cause

5. NSAIDS, combined OCP
6. Surgical management
7. NSAIDS
8. Leuprolide
9. Myomectomy

Fibroids (i.e., Leiomyoma)
-Benign tumors of the smooth muscle cells in the myometrium
-Decrease in frequency as menopause approaches
    -Tumor often requires estrogen for growth

-Related to smoking
-May cause uterine bleeding, secondary amenorrhea, dysmenorrheal, or infertility
-Diagnosis with pelvic or transvaginal ultrasound
-Treatment can be with NSAIDs and leuprolide to decrease fibroid size
-Consider surgical resection (i.e., myomectomy) if refractory to leuprolide and want to preserve fertility
-Consider hysterectomy for patients who do not desire fertility
-Leiomyosarcoma – malignant tumor of the smooth muscle cells in the myometrium which should be suspected in a patient with a mass seen on pelvic ultrasound that grows quickly within a 2 month period of time
    -Usually leiomyomas are not precursors to leiomyosarcomas

10. Leiomyosarcoma
11. CVC, culture vaginal discharge, and pelvic ultrasound

Pelvic Inflammatory Disease (PID)
-Signs and symptoms include:
    -Fever
    -Abdominal tenderness
    -Cervical motion tenderness (i.e., chandelier sign)
    -Adnexal tenderness
-Positive culture of vaginal discharge
    -PID often polymicrobial (i.e., anaerobes, aerobes)
-Leukocytosis with left shift
-Neisseria gonorrhea and Chlamydia trachomatis are most common causes of PID
    -Chlamydia being more common cause given often asymptomatic nature of infection
-Intrauterine devices (IUD) increases risk of PID
    -History of PID or concurrent sexually transmitted disease (STD) is contraindication to IUD placement
-Pelvic ultrasound helpful in workup to rule out tubo-ovarian abscess (TOA)
-Obtain pregnancy test to rule out pregnancy
-Treat either as inpatient or outpatient
    -Hospitalize if patient:
        -Noncompliant
        -HIV positive
        -Pregnant
        -Cannot tolerate food by mouth
        -If TOA present
        -Not improving on outpatient regimen
    -Otherwise, treat as outpatient
-Inpatient treatment with IV antibiotics clindamycin and gentamicin, especially for TOA
    -First step is not to drain abscess, but to start with IV antibiotics
        -Drain if antibiotics are not helpful
-Outpatient treatment with po ampicillin and doxycycline
-PID can lead to ectopic pregnancy, chronic pelvic pain, and infertility
-PID caused by gonorrhea may progress to Fitz-Hugh Curtis syndrome

12. Polymicrobial

13. Neisseria gonorrhea and Chlamydia trachomatis
14. IV or PO abx
15. Rule out tubo-ovarian abscess
16. Fitz-Hugh Curtis syndrome
17. Asherman syndrome

Secondary amenorrhea
- Amenorrhea in a woman for 6 consecutive months who has had at least 6 months of normal periods
- Causes include:
    - Asherman's syndrome
        - Scarring after delivery, endometritis, or dilatation and curettage
    - Prolactinoma
    - Hypothyroidism
    - Sheehan's syndrome
    - Anorexia
    - Excess exercise
    - Polycystic Ovarian syndrome (PCOS)
    - Antipsychotics
        - Usually begins at onset of taking medication and will stop upon stopping medication
        - Hyperprolactinemia
        - Galactorrhea
    - Pregnancy
    - Menopause
- First, check urine β-hCG level to rule out pregnancy
- Second, check prolactin and thyroid stimulating hormone (TSH)
    - If prolactin elevated, likely prolactinoma
        - Consider MRI of sella turcica to locate pituitary tumor
    - If TSH high, hypothyroidism may be the cause
        - Evaluate for hypothyroidism and consider thyroid hormone supplementation
    - If prolactin and TSH normal, then check FSH and LH
        - If low, consider Sheehan's syndrome as pituitary cause becomes more likely
        - If FSH and LH elevated, consider PCOS, premature ovarian failure, or menopause as defect likely in the ovaries
            - If FSH greater than 40, sensitive marker for menopause
                - Defined as no periods for greater than 6 months-1 year from last period
                - During perimenopause may have breakthrough bleeding
                - Periods may be irregular both at menarche and menopause
                - Complications include:
                    - Hot flashes
                    - Vaginal atrophy
                    - Osteoporosis
                    - At risk for coronary artery disease (CAD)
                - Treat vaginal atrophy with estrogen cream and give calcium supplements with or without bisphosphonates
                - Treat hot flashes with biofeedback
                    - If refractory, try clonidine

        -Consider hormone replacement therapy (HRT) for more severe, refractory symptoms
            -Major contraindication of HRT is history of deep vein thrombosis (DVT) or pulmonary embolism (PE)
-If LH to FSH ratio elevated, consider PCOS as defect likely in ovaries
-If normal, consider progesterone challenge to see if endometrium can be estrogen primed
        -If breakthrough bleeding seen, endometrium is responsive
        -If no breakthrough bleeding seen, suggests endometrium not responding to estrogen

18. Pelvic ultrasound
19. ovarian cysts bilaterally (i.e., pearl necklace sign
20. Estrogen, testosterone
21. OCP
22. Clomiphene

-Clomiphene may be used for ovulation induction in patients who desire fertility
    -Blocks the estrogen receptor in the hypothalamus decreasing negative feedback by estrogen causing increased levels of FSH and LH
    -Can get multiple eggs ovulated with use of clomiphene resulting in multiple births

23. Multiple gestations

Pediatrics

1) What is the desired lecithin to sphingomyelin ratio to assess for lung maturity?
    a. Above 1.5
    b. Above 2.5
    c. Below 1.5
    d. Below 0.5

2) If the ratio is less than 1.5, what should be administered?
    a. NSAIDS
    b. IV sphingomyelin
    c. Give mother steroids
    d. IV lecithin

A newborn baby has evidence of bloody diarrhea, fever, elevated wbc count, and a distended abdomen.

3) What is the next study of choice?
    a. Xray abdomen
    b. CT head
    c. CT chest
    d. MRI Chest

4) Xray abdomen shows gas within walls of bowels. What is likely diagnosis?
    a. Meckel's Diverticulum
    b. Volvulus
    c. Necrotizing enterocolitis
    d. Intussusception

A baby is born whose whole body is pink, pulse 120-160, grimaces when stimulated, has some flexion movements, and strong respiratory effort.

5) What is the Apgar score?
    a. 8
    b. 9
    c. 10
    d. 11

6) Does this score mean the neonate is doing well or needs resuscitation?
    a. OR now
    b. Steroids immediately
    c. Doing well, no resuscitation needed
    d. None of the above

This baby is noted on auscultation to have bowel sounds over the chest. He has had mild respiratory distress at birth.

7) What is the likely diagnosis?
   a. Volvulus
   b. Congenital Diaphragmatic Hernia
   c. Simple Pneumothorax
   d. Tension pneumothorax

8) What is treatment?
   a. Surgery
   b. Radiation
   c. Endovascular treatment
   d. None of the above

A 3-weeks-old boy presents with mother who states that he has had projectile nonbilous vomiting.

9) What physical exam finding can confirm the diagnosis?
   a. Right lower quadrant mass
   b. Olive-shaped epigastric mass may be present
   c. Left upper quadrant mass
   d. Left lower quadrant tenderness

10) What is the diagnosis?
    a. Pyloric Stenosis
    b. Achalasia
    c. Nutcracker esophagus
    d. None of the above

11) What is treatment?
    a. Radiation treatment
    b. Call pediatric surgeon and perform surgical myomectomy of pyloric sphincter
    c. Intubate immediately
    d. Nothing can be done

3-month-old female presents with bloody stools that the mother describes as very thick in nature like jelly. She has been crying quite a bit especially when her abdomen is palpated. Her abdomen is slightly distended.

12) What diagnostic test should be done next?
    a. Barium enema
    b. Ultrasound chest
    c. MRI Abdomen
    d. None of the above

13) What is this pathology?
    a. Volvulus
    b. Impaction
    c. Intussusception
    d. Appendicitis

14) What is the treatment?
    a. PO contrast
    b. Barium enema
    c. OR immediately for removal mass
    d. None of the above

A newborn child is found to have intestines herniating out of abdomen to the right of the umbilicus. There is no covering of intestines

15) What is diagnosis?
    a. Omphalocele
    b. Gastroschisis
    c. Achalasia
    d. Malrotation if the gut

16) What is treatment?
    a. Surgery
    b. Radiation
    c. Observation only
    d. Surgery then radiation

1-year-old female presents with mother with bloody stool. On abdominal imaging, a outpouching is noted 2 feet from the ileocecal valve. It is described as 2 inches long. It affects 2 % of the population.

17) What is the diagnosis?
    a. Volvulus
    b. Malrotation of the gut
    c. Meconium aspiration
    d. Meckels diverticulum

18) What complication can develop from this?
    a. Intussusception and volvulus
    b. Achalasia
    c. Gastroschisis
    d. Omphalocele

19) What is treatment?
    a. Surgery

b. Radiation
c. Chemotherapy
d. Focused steroid injection

Neonate has note passed meconium in 72 hours. Barium enema is done showing proximal end of bowel dilated and distal end bowel thin.

20) What can be done for definitive diagnosis?
a. Rectal biopsy
b. Resection of entire colon
c. MRI Rectum
d. CT rectum

21) What would be seen on biopsy?
a. Overactive neurons in wall intestine
b. Loss neurons in intestinal wall
c. Increase size in intestinal cells
d. None of the above

22) What is the diagnosis and treatment?
a. Hirschsprungs Disease, surgical resection and reanastomosis
b. Hirschsprungs Disease, radiation and surgery
c. Hirschsprungs Disease, radiation and chemotherapy
d. None of the above

A newborn Caucasian male who is blonde with blue eyes is found on prenatal workup to have PKU.

23) What is the inheritance pattern?
a. Autosomal recessive
b. Autosomal dominant
c. X linked recessive
d. None of the above

24) What complication can happen if not treated?
a. Diabetes
b. MR
c. Hypothyroidism
d. Niemann-Pick Disease

25) What is treatment of choice prior to conception for mother with PKU?
a. Mother placed on phe-free diet
b. Mother placed on Tyrosine rich diet
c. No valine and isoleucine for mother
d. None of the above

A newborn is found to have a cherry red spot on macular with MR. He is of Jewish descent.

26) What is the likely disease?
    a. Niemann-Pick Disease
    b. Tay Sachs
    c. DiGeorge Syndrome
    d. None of the above

27) What is the deficiency?
    a. Lecithin
    b. Sphingomyelinase
    c. Hexoseaminidase
    d. None of the above

Answers: Pediatrics

1) Above 1.5

Prematurity
-Born before 37 weeks of gestation
-Increases risk for:
    -Lung immaturity
        -May also be caused by oligohydramnios
            -Babies swallow amniotic fluid which helps in lung development and prevents lung hypoplasia
        -Lecithin to sphingomyelin ratio should be greater than 1.5 to 2 for lung maturity
        -Give mother corticosteroids to indirectly stimulate babies' surfactant-releasing cells to produce and release surfactant in utero to help their lungs mature
    -Bronchodysplasia
    -Retinopathy of prematurity
        -Worsened by oxygen delivery via incubator
    -Necrotizing enterocolitis
        -Bloody diarrhea
        -Fever
        -Leukocytosis
        -X-ray of abdomen shows gas within the walls of the bowels (i.e., pneumatosis intestinalis)

2) Give mother steroids
3) Xray of abdomen
4) Necrotizing enterocolitis
5) 8

APGAR Scoring
-Assign score at 1 and 5 minutes
-Appearance
    -0 – whole body is blue
    -1 – extremities are blue
    -2 – whole body is pink
-Pulse
    -0 – less than 100
    -1 – 100-120
    -2 – 120-160
-Grimace
    -0 – no response
    -1 – grimace when stimulated
    -2 – cry when stimulated
-Activity
    -0 – no activity
    -1 – some flexion
    -2 – active movement
-Respiratory Effort
    -0 – absent effort

        -1 – weak effort
        -2 – strong effort
-Score of 8-10 – neonate doing well with no need for resuscitation
-Score of 5-7 – possible need for resuscitation and continue to monitor
-Score of 0-3 – requires immediate resuscitation

6) doing well, no resuscitation needed
7) Congenital Diaphragmatic Hernia
8) surgery
9) Olive-shaped epigastric mass may be present
10) Pyloric stenosis
11) Call pediatric surgeon and perform surgical myomectomy of pyloric sphincter

12) Barium enema
13) Intussusception

Intussusception
-Segment of bowel invaginates into another segment of bowel
-Signs and symptoms include:
        -"Currant jelly," bloody stools
        -Colic
        -Inability to pass meconium
-Diagnose and treat with barium enema

14) Barium enema
15) Gastroschisis

Omphalocele versus Gastroschisis
-Omphalocele
        -Intestines herniate through midline
        -Intestines covered with umbilical sac
        -Worse than gastroschisis given its association with other congenital problems
        -Nonemergent surgical correction
-Gastroschisis
        -Intestines herniated out to the right of the umbilicus due to defect in abdominal wall
        -No umbilical or peritoneal covering of intestines
        -Treatment is surgery

16) surgery
17) Meckels diverticulum

Meckel's Diverticulum
-Persistence of the vitelline duct or yolk sac
-"Rule of 2's":
        -2% of the population
        -Presents in first 2 years of life
        -Located 2 feet from the ileocecal valve

            -2 inches long
            -2 types of epithelia (i.e., gastric and pancreatic)
-Complications include:
        -Lower gastrointestinal (LGI) bleeding
        -Intussusception
        -Volvulus
-Monitor hemodynamics and support appropriately with fluids
-May diagnose with Meckel's scan (i.e., a form of radionuclide scan)
-Treatment is surgical resection

    18) Intussusception and volvulus
    19) surgery
    20) Rectal biopsy
    21) Loss of neurons in intestinal wall

Hirschsprung's Disease
-Caused by loss of neurons in the intestinal wall
-History consistent with patient who has not passed meconium in 24-48 hours
-Proximal end dilated, distal end thin
-Barium enema shows proximal end of bowel dilated and distal end of bowel thin
-Diagnose definitively with rectal biopsy which shows lack of nerve plexuses
-Treatment is removal of segment of bowel with loss of neurons with reanastomosis

    22) Hirschsprungs Disease, surgical resection and reanastomosis
    23) Autosomal recessive
    24) MR
    25) mother placed on phe-free diet
    26) Niemann-Pick Disease

Niemann-Pick Disease
-Autosomal recessive
-Deficiency of sphingomyelinase
        -Causes buildup of sphingomyelins
-More common in Ashkenazi Jews
-Cherry red spot on macula
-Mental retardation

    27) Sphingomyelinase

Kaloostian MD- 144

Trauma and Post-Operative Management

27-year-old male s/p lumbar surgery yesterday has a fever of 39.5. He is otherwise doing well.

1) What is the most common cause for this fever?
    a. Atelectasis
    b. Pneumonia
    c. UTI
    d. Sepsis

2) What is recommended treatment for this?
    a. DC foley
    b. Use of incentive spirometer, ambulation
    c. Antibiotics
    d. None of the above

On post-operative day 10, he continues to have fevers. His wbc count is elevated. He has no Shortness of breath. His wound has induration.

3) What is most likely diagnosis?
    a. Wound infection
    b. UTI
    c. Sepsis
    d. Pneumothorax

A 45-year-old male s/p surgery has low BP, elevated CVP, elevated PCWP, and elevated SVR.

4) What type of shock is this?
    a. Cardiogenic
    b. Neurogenic
    c. Allergic
    d. Septic

5) What is the treatment?
    a. IVF, pressors
    b. Steroids
    c. NSAIDS
    d. No treatment available

6) What pressor is best for cardiogenic shock?
    a. Dobutamine
    b. Dopamine
    c. Neosynephrine
    d. Epinephrine

34-year-old male s/p surgery has decreased CVP, increased Ejection fraction, decreased PCWP, and low SVR.

7) What type of shock is this?
    a. Neurogenic
    b. Anaphylactic
    c. Cardiogenic
    d. Nephrogenic

8) What is treatment?
    a. IVF, pressors
    b. Metoprolol
    c. Steroids
    d. No treatment available

56-year-old male s/p surgery has lob BP, low VCP, elevated ejection fraction, high cardiac output, low PCWP, and low SVR.

9) What type of shock is this?
    a. Septic shock
    b. Cardiogenic
    c. Neurogenic
    d. Anaphylactic

10) What is treatment?
    a. Pan culture, then IV Abx, IVF, pressors
    b. Pan Culture then Intrathecal antibiotics
    c. Antibiotics only
    d. IVF and pressors only

24-year-old female s/p surgery has lob BP, low CVP, normal EF, low PCWP, and elevated SVR

11) What type of shock is this?
    a. Cardiogenic
    b. Anaphylactic
    c. Hypovolemic shock
    d. Septic

12) What is treatment?
    a. Pressors only
    b. IVF, pressors, blood transfusion
    c. Blood transfusion only
    d. None of the above

A 34-year-old male hiker presents with lob BP, low CVP, high EF, low PCWP, low SVR and warm skin

13) What type of shock is this?
    a. Neurogenic
    b. Cardiogenic
    c. Septic
    d. Anaphylactic

14) What is treatment?
    a. Epinephrine
    b. Steroids
    c. Norepinephrine
    d. Dobutamine

15) What emergent airway protection procedure should be done on field to obtain airway?
    a. Cricothyroidotomy
    b. Tracheostomy
    c. Hyaloid opening
    d. Hyalo-Thyroid opening

A patient is hit by a car. He has no eye opening, is intubated, and is flexor posturing.

16) What is this GCS?
    a. 5T
    b. 4T
    c. 3T
    d. 6T

A 24-year-old male is stabbed in his neck. He is stabbed below his cricoid.

17) What zone is this in?
    a. 1
    b. 2
    c. 3
    d. 4

18) What is next treatment?
    a. Angiography
    b. Balloon dilation
    c. MRI Brain
    d. Ultrasound neck

19) What are the contraindications to placement of foley?
    a. Urethral bleeding, scrotal hematoma, and high ballotable prostate
    b. Urethral bleeding only

    c. Scrotal hematoma only
    d. Urethral bleeding and scrotal hematoma only

20) What is the treatment for iron toxicity?
    a. Lead
    b. Niacin
    c. Naloxone
    d. Deferoxamine

21) What is the treatment of benzodiazepine overdose?
    a. Amiodarone
    b. Naloxone
    c. Flumazenil
    d. None of the above

22) What is treatment of methanol poisoning?
    a. Ethanol
    b. EDTA
    c. BAL
    d. Defuroxime

23) What is used to reverse heparin?
    a. Protamine sulfate
    b. Coumadin
    c. Naltrexone
    d. Vitamin K

24) What is used to reverse Organophosphate poisoning?
    a. Atropine only
    b. Pralidoxime only
    c. Atropine and pralidoxime
    d. None of the above

Kaloostian MD- 149

Answers: Trauma and Post-Operative Management

1) Atelectasis

Post-Operative Fevers

POD 1 → most common cause is atelectasis; incentive spirometer will help prevent; ambulation is also helpful
POD 3 → UTIs, pneumonia; diagnosis with CHEST X-RAY, sputum gram stain/cx, UA, blood cx
POD 7 → pulmonary embolism, sinusitis (secondary to long-term nasogastric tube placement); diagnosis pulmonary embolism clinically by swollen/tender extremity and with Doppler ultrasound; consider V/Q scan or CT spiral Angio depending on clinical suspicion; prevent with SCDs and DVT prophylaxis
POD 10 → wound infection, decubitus ulcer, cellulitis, intrabdominal abscess; diagnosis wound infection with wound cx; prevent decubitus ulcers with q2 hr turning and soft mattress; diagnosis cellulitis clinically; if considering intrabdominal abscess, CT with IV contrast may help visualize
Drugs (i.e., bactrim, phenyotin) → can cause fever on any POD; treatment is to stop drug that may be causing it if not necessary for the pt

2) Use of incentive spirometer, ambulation
3) wound infection
4) Cardiogenic

Shock

Cardiogenic Shock
-secondary to decreased EF
-low BP
-elevated CVP due to back flow as blood not leaving the heart
-elevated pulmonary capillary wedge pressure (PCWP)
-elevated SVR
-pale, cool clammy skin
-e.g., CHF and its various causes (MI, endocarditis, emboli, HTN, cardiomyopathy (viral))
-check cardac enzymes, do Echo, give fluids, pressors if necessary (dobutamine first-line given its inotropic effect)

Neurogenic Shock
-body cannot vasoconstrict as compensation for low BP secondary to nerve damage
-decreased CVP secondary to venodilation
-increased to normal EF
-decreased PCWP
-low SVR
-warm skin
-e.g., toxins, diabetes, neurologic disorders, trauma
-give fluids, pressors if necessary

Systemic Inflammatory Response Syndrome (SIRS)

-require 2 or more of the following:
- fever greater than 38 C or temp less than 36 C
- tachycardia
- increased WBC
- tachypnea

Sepsis
-blood cultures are positive and SIRS present

Septic Shock
-vasodilation secondary to sepsis/bacteremia
-pts can become septic from UTIs/pyelonephritis, pneumonias, skin infections)
-low BP
-normal to low CVP secondary to venodilation
-elevated EF (high ouput CHF)
-low PCWP
-low SVR
-warm skin
-first, pan-culture; then, give IV antibiotics, start fluids (most importantly), consider pressors (start with dobutamine)

Hypovolemic Shock
-low blood volume
-e.g., trauma (hemorrhage), burns
-low BP, tachycardic
-low CVP secondary to blood loss
-normal EF
-low PCWP
-elevated SVR
-pale, cool clammy skin
-give fluids (beware of pressors initially, normalize CVP first before considering pressors); consider RBC tranfusions as needed (follow serial hematocrits)
-hypovolemic shock is classified depending on amount of blood loss
- Class 4 if greater than 40 % loss (pts cannot compensate anymore)

Anaphylactic Shock
-secondary to severe allergic reaction to food, beestings, etc…causing systemic vasodilation
-low BP
-low CVP secondary to venodilation
-high EF
-low PCWP
-low SVR
-warm skin
-treat with 1:1000 epinephrine; intubate if pt still in respiratory distress

5) IVF, pressors
6) Dobutamine
7) Neurogenic
8) IVF, pressors
9) Septic shock

10) Pan culture, then IV Abx, IVF, pressors
11) Hypovolemic shock
12) IVF, pressors, blood transfusion
13) Anaphylactic
14) epinephrine
15) Cricothyroidotomy

If unable to attain airway after 3 attempts or in the case of facial trauma (i.e., facial fractures), attempt surgical airway in adults/pediatrics
-cricothyroidotomy – short-term airway in emergency setting
-tracheostomy – long-term airway management for pts who have been intubated for longer than 2 wks; they must be transitioned to tracheostomy so as to prevent laryngotracheomalacia; requires OR usually, and is more convenient for the pt

16) 5T

Glasgow Coma Scale (GCS)
-conducted during D (disability) portion of primary survey
-conducted on every trauma patient to determine severity of neurological injury
-scale ranges from 3-15
-based on eye opening, verbal, and motor ability
-Eye opening
- -1- no spontaneous movement
- -2- opens eyes to pain
- -3- opens eyes to voice
- -4- opens eyes to spontaneously

-Verbal
- -1- no sounds uttered, intubated
- -2- incomprehensible speech
- -3- words discernible, inappropriate response
- -4- confused conversation, but able to answer
- -5- fully oriented

-Motor
- -1 – No movement to painful stimuli
- -2 – extensor response to painful stimuli
- -3 – flexor response to painful stimuli
- -4 – withdraws to painful stimuli
- -5 – localizes to painful stimuli
- -6 – moves extremities to command

-if trached/intubated, count as T for verbal (count as 1 in overall sum)
-GCS less than 7 requires immediate intubation

17) 1
18) Angiography
19) Urethral bleeding, scrotal hematoma, and high ballotable prostate

Contraindications to Foley Placement
-Bleeding at urethral meatus
-Scrotal hematoma

-High-ballotable prostate
-need to conduct retrograde cystourethrogram to determine if urethra has been ruptured – if try to apply foley in such a pt, placing the foley may turn a minimally ruptured urethra into a total rupture

20) Deferoxamine
21) Flumazenil
22) Ethanol
23) Protamine sulfate
24) Atropine and pralidoxime

Dermatology

24-year-old female has noted that she has a vesicular rash around the area of her earrings and watch.

1) What is the likely diagnosis?
    a. Atopic dermatitis
    b. Acne
    c. Eczema
    d. Contact dermatitis

2) What type of hypersensitivity reaction is this?
    a. 4
    b. 3
    c. 2
    d. 1

3) What is treatment?
    a. Antibiotics
    b. Avoid allergen, topical steroids
    c. NSAIDS
    d. None of the above

56-year-old female has itchy erythematous rash on flexural surfaces of her body. She has eosinophilia in her blood, and has a past medical history significant for asthma.

4) What is the most likely diagnosis?
    a. Atopic dermatitis (Eczema)
    b. Acne
    c. Syphilis
    d. Parvovirus B 19 infection

5) What is the treatment?
    a. NSAIDS only
    b. Cool compresses
    c. Avoidance trigger, topical steroids
    d. Antibiotics

27-year-old female has multiple oval lesions on her skin that have a silvery appearance on extensor surfaces of her body. She has tried to remove a scale but this started bleeding.

6) What is the diagnosis?
    a. Psoriasis
    b. Rosacea
    c. Acne vulgaris
    d. Eczema

7) What is the term for the bleeding after removing scale?
    a. Auspitz sign
    b. Bleeding sign
    c. Purple Sign
    d. Scab sign

8) What is treatment?
    a. NSAIDS only
    b. Radiation only
    c. Psoralen only
    d. Psoralen + UV radiation or topical steroids

9) What other pathology can be seen in these patients?
    a. Brian tumors
    b. Nail pitting, sausage digits, and arthritis
    c. Cardiac hypertrophy
    d. None of the above

A 56-year-old male has multiple stuck on appearing lesions on his neck and back.

10) What is diagnosis?
    a. Seborrheic keratosis
    b. Actinic keratosis
    c. Basal cell cancer
    d. Melanoma

11) With what cancer are they associated with?
    a. Gastric
    b. Lymphoma
    c. Kidney
    d. Liver

12) What is treatment?
    a. Radiation
    b. Cryotherapy
    c. Chemotherapy
    d. None of the above

56-year-old male business executive is worried about his face. He has noticed increased redness with small tiny blood vessels on his cheek and nose. His nose has also increased in size significantly.

13) What is the diagnosis?

a. Rosacea
b. Eczema
c. Acne vulgaris
d. None of the above

14) What is the term given to enlargement of nose in this condition?
    a. Rhinophyma
    b. Syringomyelia
    c. Syringobulbia
    d. None of the above

15) What is treatment?
    a. Reduce stress only
    b. Reduce stress and alcohol only
    c. Antibiotics only
    d. Reduce stress, reduce alcohol and make-up, oral tetracycline

A 45-year-old male in given an antibiotic. He immediately develops painful ulcers and lesions throughout his mucous membranes and lips.

16) What is the diagnosis?
    a. Simple allergic reaction
    b. Stevens Johnson syndrome
    c. Rosacea reaction
    d. Severe Burn reaction

17) What is treatment?
    a. IVF resuscitation, stop medication, epi
    b. Cool compresses
    c. Dantrolene
    d. No treatment available

A popular singer is embarrassed as he has noticed multiple patches of hypopigmented regions throughout his body. He also has hypothyroidism and type 1 DM.

18) What is the diagnosis?
    a. Vitiligo
    b. Acne vulgaris
    c. Melanoma
    d. Lentigo

19) With what HLA is this associated with?
    a. HLA DR ¾
    b. HLA DR 5
    c. HLA DR 9

d. HLA DR 1

20) What is treatment?
    a. Steroids only
    b. Steroids + UV radiation, sun protection
    c. UV radiation only
    d. Steroids + avoid sun only

78-year-old male Caucasian farmer presents with multiple areas of pearly lesions on his skin with rolled up borders.

21) What is next step in diagnosis?
    a. MRI Head
    b. CT head and chest
    c. Punch biopsy
    d. Laser Removal

Biopsy shows islands of epithelium with peripheral palisading.

22) What is diagnosis?
    a. Basal cell carcinoma
    b. Melanoma
    c. Actinic Keratosis
    d. Rosacea

23) What is treatment?
    a. Radiation
    b. Chemotherapy
    c. Surgical resection
    d. Accutane

Answers: Dermatology

1) Contact dermatitis

Contact Dermatitis
-Type 4 hypersensitivity reaction
-Langerhan's cells (i.e., dendritic cells) initially carry antigen involved to T-helper cells which further promote the reaction
-Most common cause is nickel
    -Poison ivy and cosmetics may also cause
-Lesions can be vesicular and are usually linear or take the shape of the object causing the reaction (i.e., watch, sandal)
-Clinical diagnosis
-Treatment involves avoiding allergen
-Use topical corticosteroids if necessary

2) 4

Type of Hypersensitivity Reactions
-Type I Hypersensitivity Reaction (i.e., allergic mediated, IgE mediated)
    -Immediate reaction – urticaria, hives
        -Once patient is exposed to allergen for the first time, they produce IgEs that bind to mast cells
        -Once exposed to the allergen a second time, patients develop symptoms as allergen binds to IgE on mast cells resulting in crosslinking that stimulates the release of histamine that causes vasodilation, edema, redness, and urticaria
    -Self-limited with removal of agent
    -Seen in bee stings, food allergies, and anaphylaxis
-Type II Hypersensitivity Reaction (i.e., cytotoxic, antibody mediated)
    -Antibodies cause the damage
    -Seen in Goodpasture's syndrome, Graves disease, Autoimmune hemolytic anemia, Rh incompatibility, Rheumatic fever
-Type III Hypersensitivity Reaction (i.e., complement mediated, immune complex mediated)
    -Immune complexes cause the damage
    -Seen in Serum sickness, Arthus reaction, Lupus, Rheumatoid arthritis, and Post-streptococcal glomerulonephritis
        -Serum sickness
            -Occurs 8-12 days after exposure to agent
            -Secondary to injection of foreign proteins like horse serum
            -Causes local reaction
        -Arthus reaction
            -More acute of a reaction than serum sickness
            -Local antigen-antibody mediated reaction that activates complement causing inflammatory cells to enter vessel walls leading hemorrhagic necrosis
-Type IV Hypersensitivity Reaction (i.e., delayed type hypersensitivity, T cell mediated)
    -Takes 48-72 hours to develop reaction
    -Only hypersensitivity reaction that is not transferable by serum given antibodies are not involved

-Seen in Tuberculosis skin test, Hashimoto's thyroiditis, Contact dermatitis, and Transplant rejection

3) Avoid allergen, topical steroids
4) Atopic dermatitis(eczema)
5) Avoidance trigger, topical steroids
6) Psoriasis

Psoriasis
-Caused by increased division and proliferation of cells in the epidermis (i.e., epidermal hyperplasia)
-Classic silvery scales appearance
-Usually localized on extensor surfaces (i.e., elbows or knees)
-Complications include:
    -Nail pitting
    -Sausage digits
    -Arthritis (i.e., HLA B27 positive, RF negative)
-Positive Auspitz sign – if remove scale or at site of trauma, area starts bleeding
-Classic histology reveals Munro's abscesses which are neutrophilic infiltrates in the stratum corneum
    -Histology usually not required for diagnosis
-Clinical diagnosis
-Treat with psoralen plus UV A radiation or topical corticosteroids

7) Auspitz sign
8) Psoralen + UV radiation or topical steroids
9) Nail pitting, sausage digits, and arthritis
10) Seborrheic keratosis

Seborrheic Keratosis
-Has stuck on appearance
-Warty-like growth
-Noncancerous
-Treat with cryotherapy
-Associated with gastric cancer

11) Gastric cancer
12) Cryotherapy
13) Rosacea

Rosacea
-Skin appears erythematous with telangiectasias
-More common in Caucasian women
-Usually affects the face
-If affects the nose, can get rhinophyma (i.e., large nose)
-Clinical diagnosis

-Treatment is prevention by reducing stress, alcohol, and heavy use of cosmetics and other irritating skin products
-Add oral antibiotics (i.e., tetracycline or metronidazole)

14) Rhinophyma
15) Reduce stress, reduce alcohol and make-up, oral tetracycline
16) Stevens-Johnson Syndrome
17) IVF resuscitation, stop medication, epi
18) Vitiligo
19) HLA DR ¾
20) Steroids + UV radiation, sun protection
21) Punch biopsy
22) Basal cell carcinoma
23) Surgical resection

Epidemiology

1) The false negative rate in a population is 1.0. What is the sensitivity?
    a. 0
    b. 1
    c. 2
    d. 3

2) The false positive rate in a population is 0.5. What is the specificity?
    a. 0.5
    b. 1
    c. 2
    d. 3

Total number of patients without the disease is 100 and of those only 50 have a negative test result. Total number of people with negative test results is 300.

3) What is negative predictive value?
    a. 1
    b. 0.5
    c. 0.7
    d. 0.9

4) Power is calculated as what?
    a. 1-Beta
    b. 1-alpha
    c. 2- Beta
    d. 2-Alpha

5) Beta in the above equation is the probability of making what type of error?
    a. Type I
    b. Type II
    c. Type III
    d. None of the above

6) What error is the probability of stating there is no difference even though a difference exists?
    a. Type II-Power
    b. Type I -Power
    c. Both Type I and II
    d. None of the above

7) What validity refers to the ability to generalize from a sample to its greater population?
    a. Internal validity

b. Superficial Validity
   c. Power
   d. External validity

8) What error is the probability of stating there is a difference when a difference does not exist?
   a. Type I
   b. Type II
   c. Type III
   d. Type IV

9) What is the typical alpha level used to denote significance in a study?
   a. Less than 0.005
   b. Less than 0.05
   c. Less than 0.5
   d. None of the above

10) What is the confidence interval within one standard deviation of the mean?
    a. 67%
    b. 75%
    c. 95%
    d. 98%

11) In a population, the attributable risk is 0.5. What is the Number needed to treat?
    a. 0.3
    b. 0.5
    c. 0.6
    d. 0.7

12) What study type involves a report on one person or event and lacks power?
    a. Case Study
    b. Case Report
    c. Observational
    d. Randomized controlled Trial

13) What study type follows groups over time either prospective or retrospective?
    a. Cohort study
    b. Case Report
    c. Randomized controlled trial
    d. Meta-Analysis

14) What study type follows two groups and looks back in their histories to see if they were exposed to a risk factor of interest and that is affected by recall bias?
    a. Case Report

b. Observational Study
c. Meta-Analysis
d. Case-Control Study

15) What study type is the highest level of evidence that looks at multiple different studies together showing significant results?
    a. MetaAnalysis
    b. Cohort Study
    c. Case Report
    d. Randomized controlled trial

16) What type of bias refers to the situation when a screening test tends to find a disproportionate number of diseases that are slowly progressive compared to those that are rapidly progressive?
    a. Observer bias
    b. Link Bias
    c. Length bias
    d. Non randomization

17) What type of bias refers to the situation when refers to the situation when a screening test finds the disease earlier on in its progression, and thus we begin to overestimate the incidence of the disease
    a. Lead time bias
    b. Observer Bias
    c. Interlinker Bias
    d. None of the above

18) Mumps vaccine is an example of what type of vaccine?
    a. Live attenuated
    b. Dead vaccine
    c. Live non-attenuated vaccine
    d. None of the above

19) A contraindication to live attenuated vaccines is:
    a. Pregnancy
    b. Liver disease
    c. Kidney disease
    d. Pneumonia

20) What new vaccine is given as 3 doses to women above the age of 9 to protect against cervical cancer, but does not protect against all strains?
    a. Colon cancer
    b. Prostate cancer
    c. Kidney cancer
    d. HPV vaccine

Answers: Epidemiology

1) 0

Sensitivity
-The number of people with the disease and a positive test result divided by the total number of people with the disease
-Also equal to: 1 – False-negative rate

Specificity
-The number of people without the disease and a negative test result divided by the total number of people without the disease
-Also equal to: 1 – False-positive rate

2) 0.5
3) 0.5
4) 1-Beta

Power
-Equivalent to $1 - \beta$
-β is the probability of making a type II error
    -Type II error – is the probability of stating there is no difference between the groups under study even though a difference exists in the true population
    -By increasing power, you decrease the probability of making a type II error, and thus are more likely to find a true difference in the population if one exists
-Relates to external validity
    -External validity – refers to the ability to generalize from a sample to its greater population
-The greater the power, the greater the sample size, and thus the better we are able to generalize from a sample to the population

5) Type II
6) Type II-Power
7) External validity
8) Type I
9) less than 0.05
10) 67%
11) 0.5
12) Case Study

Case Report
-Report on one person or event
-Lacks power as it involves only one individual or event
-Effective if individual or event under study is extremely rare
-Case Series – similar to case report, although it involves the study of more than one person or event

Cross-Sectional Study
-Analyzes the relationship between two variables at one point in time

-Can calculate a correlation between the two variables
-Can calculate prevalence
-No randomization involved
-Many confounds
-Cannot evaluate for causation

Cohort Study
-Follows two groups over time, one with a risk factor and another without, to determine if an outcome of interest develops
-Can be prospective (i.e., when you follow the groups in time) or retrospective (i.e., where you look back in historical records to identify a cohort and look back at their histories up to the present time)
    -Retrospective studies are more prone to bias than prospective studies, specifically due to historical or recall bias
-Can calculate incidence and prevalence
-Can calculate relative risk and attributable risk
-Good for studying common diseases
-Not good for studying rare diseases
-Good for studying common risk factors
-Expensive
-Study takes a long time
-No randomization involved
-Can have people who leave the study
    -Such patients should be recovered in an intention-to-treat analysis

Case-Control Study
-Follows two groups, one with the outcome of interest and the other without, and looks back in their histories to see if they were exposed to a risk factor of interest
-Can calculate an odds ratio
-Good for studying rare diseases
-Not good for studying rare risk factors or exposures
-Affected by historical or recall bias as participants might forget their histories

Randomized Clinical Trials (RCTs)
-Also called experiments
-Involves randomization into groups
-Control group versus treatment group
-Can evaluate for causation
-Harbors less confounding variables than the previous study designs

Metanalysis
-Highest level of evidence
-Groups multiple different studies together studying the same relationship to increase the power to obtain a statistically significant result even though the individual studies were not statistically significant
-Can show that a clinically significant result is statistically significant by increasing the power
-Only as good as individual studies that make it up
    -If the individual studies are biased, the metanalysis will also be biased

   13) Cohort Study
   14) Case-Control Study

15) Meta-Analysis
16) Length bias

Bias
-Any method that causes there to be a systemic difference in the way results are obtained, measured, etc…
-Measurement bias – refers to differences in measurement between two groups
-Sample or Selection bias – refers to two groups who are different with respect to variables that are not specifically being compared in the study
-Historical or Recall bias – refers to the scenario where participants forget information from their past that is relevant to the study
-Lead-time bias – refers to the situation when a screening test finds the disease earlier on in its progression, and thus we begin to overestimate the incidence of the disease
    -It gives the impression of prolonged survival
-Length bias – refers to the situation when a screening test tends to find a disproportionate number of diseases that are slowly progressive compared to those that are rapidly progressive
    -We tend to overestimate the benefit of the screen

17) Lead time bias
18) Live attenuated

Live Attenuated Vaccines
-Measles vaccine
    -Contains neomycin
-Mumps vaccine
    -Contains neomycin
-Rubella vaccine
    -Contains neomycin
-Varicella vaccine
    -Contains neomycin
-Zoster vaccine (i.e., Zostavax)
    -Contains neomycin
-Yellow fever vaccine
-Oral polio vaccine (i.e., Sabin)
-Hepatitis A vaccine
-Contraindications include:
    -Pregnancy
    -Immunosuppression
    -HIV
    -Previous anaphylaxis to any of the components of the vaccine (i.e., neomycin)
-Mild fever, being sick at the time of the vaccine, or mild allergy including skin rash are not contraindications to vaccination

Inactivated Vaccines
-Hepatitis B vaccine
    -Recombinant vaccine
    -Only vaccine given at birth
-Pneumococcal vaccine (i.e., Pneumovax)
    -Valent vaccine

-Neisseria meningitides vaccine (i.e., Menactra)
    -Valent vaccine
-DTaP vaccine (i.e., diphtheria, tetanus, pertussis)

   19) Pregnancy
   20) HPV vaccine

Summary:

We hope that after studying and testing oneself in this textbook that one has learned the information we feel is commonly tested on the USMLE board examination. We have included the most high yield pathological conditions and have used the information presented in the text book to prepare appropriate USMLE style questions. We remember the process of going through USMLE 1 to 3, and then continuing with our Specialty board examinations. After much attention and review, we feel that we have come to a better understanding of how to approach these examinations and how best to study the immense amount of material that is expected to be understood. This is the perspective we have used when writing these texts. It is with this in mind that we hope these textbooks can be utilized to effectively do well on USMLE examinations. Good luck on your boards and in the future. Please feel free to contact any of us with any questions or concerns. Thank you.

Hippocratic Oath

I swear by all that I hold most sacred
That I will keep this enduring oath:

To the best of my ability and judgement I will
Practice the Art only for the benefit of my patients.

Whatever houses I may visit, I will enter only
To help the sick or to prevent illness,
Never to inflict harm, injustice, or suffering.

I will lead my life and practice the Art
Conscientiously and with honor.

Whatever I may see or hear in the practice of the Art
Or even outside of it that should not be spread abroad
I will keep in solemn confidence.

I will be just and generous to those who taught me
The Art, to my colleagues in the Art, and to
Those who desire to learn it.

May happiness and the physician's good repute be
Granted me while I keep this sacred Oath inviolate.

www.ingramcontent.com/pod-product-compliance
Lightning Source LLC
Chambersburg PA
CBHW080913170526

45158CB00008B/2092